BASIC PRINCIPLES OF ELECTROSTATICS

BASIC PRINCIPLES OF ELECTROSTATICS

R.K. Taneja

CYBER TECH PUBLICATIONS
4264/3, Ansari Road, Daryaganj, New Delhi-110002 (India)
Ph.: 011-23244078, 011-43559448 Fax: 011-23280028
E-Mail: cyberpublicationsdelhi@yahoo.com
Website: www.cybertechpublications.com

BASIC PRINCIPLES OF ELECTROSTATICS

R.K. Taneja

First Edition 2010

Published by :

G.S. Rawat for **Cyber Tech Publications**
4264/3, Ansari Road, Daryaganj, New Delhi-110002 (India)
Ph.: 011-23244078, 011-43559448 Fax: 011-23280028
E-Mail: cyberpublicationsdelhi@yahoo.com
Website: www.cybertechpublications.com

PREFACE

The books has been specially written to the requirements syllabus of the students of all Indian Universities. Sufficient problems has also been selected from various universities examination papers. and the author shall be thankful to the readers who point out errors and omissions which inspite of all care might have been there.

The chapter of this book contains complete theory and fairly large number of some examples. The subject matter has been discussed in such a simple way that the students will find no difficulty to understand it. The proofs of various theorems and examples has been given with minute details.

The author will feel highly obliged to all students for their constructive suggestions and healthy criticism of the book which will go a long way in the imporovement of the text. If the book serve the purpose for which it is meant and improvement of the book are always welcome.

—Author

PREFACE

Contents

Pages

Preface

1. **Chemical Effect of Current** 1

Introduction, First Law of Faraday, Second Law Faraday, Faraday Constant, Relation between Faraday's Constant, Avogadro's number and Fundamental Charge, Resistance of Electric, Resistivity or Specific Resistance, Resistance of Pure Metals, Resistance of Alloys, Resistance of Semiconductors, Resistance of Electrolytes, Resistance of Superconductors, Internal Resistance of a Cell, Kirchoff's Laws, Combinations of Resistances, Wheatstone's Bridge, Heating Effect of Current and Thermo-Electricity, Shunt, Ammeter, Voltmeter and Branching of Current, Current Density, Potential Difference, Solved Examples.

2. **Attracted Disc Machines and Electrometers** 70

Introduction, Physical Meaning of Polarisation of Dielectric, Construction of Electrometers, Procedure for Measuring the Potential Difference, Heterostatic Use, Idiostatic Use, Measurement of Capacity, Measurement of Feeble (Ionisation) Currents, Measurement of Dielectric Constant, Measurement of Dielectric Constant Using Attracted Disc Electrometer, Solved Examples, Exercises.

3. **Dielectric Susceptibility and Permittivity** 97

Introduction, Dipole Approximation to Polarised Dielectrics, Boundary Conditions at the Surface Separating two Substances, Boundary Conditions at the Surface of a Charged Conductor, Faraday Tubes (Unit Tubes of Induction), A Cylindrical Capacitor, Spherical Capacitors,

Guard Ring Capacitor, Energy Stored in a Capacitor, Force of Attraction Between Capacitors Plates, The Dielectric Constant, Classification of Dielectric on the Basis of Molecular Structure (Polar and Non Polar Molecules), A Dielectric in an Electrostatic Field (Polarisation of Dielectric), Capacity of a Parallel Plate Capacitor Partially Filled with Dielectric, Dielectrics and Gauss's Law, Capacity of a Parallel Plate Capacitor when Completely Filled with Dielectric, Electric Polarisation Vector P, Uniform and Non-Uniform Polarisation, Three Electric Vectors, Electronic Polarisability, Molecular Polarisability, Interfacial Polarisability, Limitations of the Clausius Mosotti Equation, Ferroelectric and Paraelectric Dielectric, Curie Temperature, Pressure at a Conductor (Plates)—Dielectric Boundary, Solved Examples, Exercises.

4. Different Series of Currents **155**

Introduction, Current Magnification, Purely Resistance Circuit, Power in Inductive Circuit, Purely Capacitative Circuit, Circuit Containing L, C and R, Choke Coil, Alternating Electric Current, Virtual or Effective Value of Sinusoidal Voltage, Important Points About Complex Numbers, Pure Resistance Circuit containing, Capacitance in Series, A.C. Circuit with Resistance, Capacitance, Solved Examples, Exercises.

5. Applicaition of Potentiometer Current **206**

Introduction, Calibration of an Ammeter and Voltmeter, Measurement of Very Small Thermo EMF, Measurement of Large Potential Difference, Determination of Internal Resistance of a Cell, Comparison of Two Low Resistances, Determination of a Given Low Resistance, Definition of Potential Difference, Current and Current Density, Current Density, Wheatstone's Bridge, Kelvin's Method for Determining the Resistance of a Galvanometer, Specific Resistance, Ohm's Law at a Point, The Relaxation Time, First Current Law of Kirchhoff's, Kirchhoff's Voltage (Second) Law or Loop Rule, Solved Examples, Exercises.

1

Chemical Effect of Current

INTRODUCTION

Faraday carried out experiments on different electrolytes and proposed following two laws of electrolysis on the basis of his observations:

FIRST LAW OF FARADAY

The mass of an element liberated at an electrode in the process of electrolysis, is directly proportional to the quantity of electricity of total electric charge passed through the electrolyte. If m kg is the mass of a substance liberated due to flow of a current of i ampere for t seconds, then

$m \propto q$ (where q represents the quantity of charge flown in t sec)

Because $q = it$, hence $m \propto it$

or $\mathbf{m = Zit}$

Important Points

(a) Z is constant for the given element, called as *Electro chemical equivalent* or *E.C.E.*

(b) If i = 1 sec, Then Z = m i.e. E.C.E. of an element is equal to the mass of the element liberated when one ampere of current is passed through the electrolyte for one sec.

(c) $Z_{Ag} = 1.118 \times 10^{-6}$ Kg/coulomb;

$Z_{Cu} = 0.329 \times 10^{-6}$ Kg/coulomb

(d) Unit of Z in SI system = Kg/coulomb

(e) $[Z] = [m]/[q] = [MA^{-1}T^{-1}]$

(f) **International ampere :** One international ampere is that value of electric current which when passed through a solution of silver nitrate for one second, liberates 1.118×10^{-6} Kg of silver.

SECOND LAW FARADAY

If same quantity of electricity is passed through different electrolytes simultaneously (by joining different voltameters in series), then masses of different substances liberated are found to be in the direct ratio of their chemical equivalents.

Note : If the atomic weight of an element be A and its valency be V, then its chemical equivalent is given by

$W = A/V$ e.g. $W_{Ag} = 107.9/1 = 107.9$,

$W_{Cu} = 63.5/2 = 31.75$

When chemical equivalent is expressed in Kg, then it is called Kg-Equivalent,

e.g. 1 kg - equivalent of Ag = 107.9 Kg.

1 kg – equivalent of Cu = 31.75 Kg.

If masses of elements liberated are m_1 and m_2 due to passage of same quantity of charge through two different electrolytes simultaneously and W_1 & W_2 be their chemical equivalents respectively, then according to Faraday's 2nd law

$$\frac{m_1}{m_2} = \frac{W_1}{W_2}$$

e.g. If same charge is allowed to pass through water, copper and silver voltameters simultaneously, then hydrogen, copper and silver liberated at the cathodes will be in the ratio 1 : 31.75 : 108 respectively i.e. in the ratio of their chemical equivalents.

FARADAY CONSTANT

According to Faraday's second law,

we know that $$\frac{m_1}{m_2} = \frac{W_1}{W_2} \qquad ...(1)$$

If Z_1 and Z_2 are the electrochemical equivalents of two elements, then according to Faraday's 1st law, we have $m_1 = Z_1$ it and $m_2 = Z_2$ it

$$\frac{m_1}{m_2} = \frac{Z_1\ it}{Z_2\ it} = \frac{Z_1}{Z_2} \qquad ...(2)$$

Comparing equation (1) & (2)

$$\frac{Z_1}{Z_2} = \frac{W_1}{W_2} \text{ or } \frac{W_1}{Z_1} = \frac{W_2}{W_2}$$

$$\text{or } \frac{W}{Z} = \text{A Constant} \qquad ...(3)$$

Because the ratio W/Z is same for all the elements, hence it is called as Faraday's constant F.

For Example :

$$\text{(i)}\quad F_{Ag} = \frac{W_{Ag}}{Z_{Ag}} = \frac{107.9\ \text{Kg}}{1.118 \times 10^{-6}\ \text{Kg/Coulomb}}$$

$$= 9.65 \times 10^7\ \text{Coulomb/Kg-Equivalent}$$

$$\text{(ii)}\quad F_{Cu} = \frac{W_{Cu}}{Z_{Cu}} = \frac{31.75\ \text{Kg}}{0.329 \times 10^{-6}\ \text{Kg/Coulomb}}$$

$$= 9.65 \times 10^7\ \text{Coulomb/Kg-equivalent}$$

Thus *Faraday's constant is equal to the quantity of charge required to liberatз one kg - equivalent of any substance by electrolysis.* Its value is 9.65×10^7 Coulomb/Kg-equivalent.

Note : Charge required to liberate 1 gm - equivalent of any substance = 96,500 coulomb = 1 Faraday.

RELATION BETWEEN FARADAY'S CONSTANT, AVOGADRO'S NUMBER AND FUNDAMENTAL CHARGE

As charge required to liberate 1 kg – equivalent of any substance by electrolysis = F

Hence, charge required to liberate 1 kg-atom of any substance of valency V = FV.

According to Avogadro's Hypothesis, 1 kg. atom of any substance consists of N (= 6.02×10^{26}) atooms. The number N is known as Avagadro's number.

Hence, charge required to deposit Natoms of any substance = FV i.e. charge required to deposit one atom of a substance = FV/N.

In the process of electrolysis, atoms liberated at the electrode reach there in the form of ions and after giving their charge to the electrode, are liberated in the form of atoms.

i.e. If, for example, a Cu^{++} ion reaches the cathode, then it is liberated after giving 2e charge to the electrode i.e. charge required to liberate one atom of valency V = Ve

Therefore, (FV/N) = Ve or $F = Ne$.

Conduction in Metals

(i) Metals are very good conductors of electricity among solids. In any atom of a substance, the electrons in the orbits nearer to nucleus are tightly bound to the nucleus but the electrons in the outer orbits are very loosely bound and can be easily detatched from the atoms. These electrons move freely in the vacant spaces between the atoms and are known as conduction electrons of Free electrons. It is these electrons which act as charge carries for conduction of electricity in metals. Because metals are found to have a very high number of these electrons, hence they are good conductors of electricity. *Silver is best conductor of electricity.*

Note : Electric conduction takes place in certain liquids and gases also but there is an important difference between electric conduction in metals and that found in liquids and gases. In metals, it is ionly the –ve charges which take part in conduction while in liquids and gases, both –ve and +ve charges are responsible for conduction of electricity.

(ii) **Drift Velocity**

(a) As the free electrons in metals behave in the same way as do the molecules of a gas, hence they are said to form *Electron gas.* Just like the molecules of a gas, these electrons are also in a state of random motion. They constantly collide with each other and go on changing their direction of motion i.e. they have no net motion in any particular direction i.e. net rate of flow of charge through any plane in the metal is zero i.e. there is no current in the metal.

(b) When we connect the ends of a metallic wire with the terminals of a battery, then an electric field is produced at every point of the wire. Due to this electric field, every free electron now experience a force and gets accelerated in a particular direction i.e. opposite to that of electric field. This leads to establishment of current in the wire.

(c) But the electrons are not accelerated indefinitely but for a very short time. After a momentary acceleration, the electron always collides with a positive ion and loses its energy obtained from battery in the form of heat energy. It is then again accelerated between two positive ions and again collides with next positive ion and so on. *Thus, applied potential difference does not give an accelerated motion to the electrons but only gives them a very small constant velocity which gets superposed on the random motion of electrons.* This small constant velocity is known as *Drift velocity.*

RESISTANCE OF ELECTRIC

(i) Whenever a potential difference is established between the ends of a conductor, an electric current starts flowing in it. The ratio of potential difference applied across the conductor and the current flowing in it is called as Electric Resistance R of the conductor. If V is the potential difference applied across the conductor and i is the current flowing in it, then electric resistance R is : $R = V/i$

(ii) Unit of resistance in SI system is ohm;

1 ohm = [1 Volt/ 1 Ampere]

(iii) 1 megha ohm = $1\ M\ \Omega = 10^6$ ohms;

1 micro ohm = $1\ \mu\ \Omega = 10^{-6}$ ohms

1 milli ohm = $1\ m\ \Omega = 10^{-3}$ ohms

(iv) Reciprocal of Electric Resistance is called electric conductance and its unit is ohm^{-1} or Mho.

(v) $$[R] = \frac{[V]}{[I]} = \frac{[W/Q]}{[I\}} = \frac{[W]}{[I]\,[Q]} = \frac{[ML^2T^{-2}]}{[A]\,[AT]}$$

$$= [ML^2T^{-3}A^{-2}]$$

Ohm's Law

(i) If the physical state of a conductor remains unchanged (i.e. temperature, material and dimensions), then the ration of the potential difference applied across its ends to the current flow in it, remains constant. i.e. If V is the potential difference applied across the ends of the conductor and i is the current flowing through it, then according to ohm's law

$$\frac{V}{i} = \text{A Constant}$$

(ii) Because, the ratio V/i is called as the electrical resistance R of the conductor, hence

$$\frac{V}{i} = \text{A Constant} = R$$

Hence ohm's law can now be expressed as – If the physical state of a conductor remains unchanged, then Resistance of the conductor remains constant, whatever may be the potential difference applied across it. In other words, a graph plotted between applied potential difference V and the current i flowing through the conductor will always be a straight line.

Note : Ohm's law is valid for metallic conductors only.

(iii) **Proof of Ohm's Law**

Let us consider a conductor of length/and cross sectional area A. Suppose a current i flows through the conductor when a potential difference V is applied across its ends. If n is the number of free electrons per unit volume and v_d is the drift velocity of electrons, then we know that

$$i = neAv_d \qquad ...(1)$$

Because V is potential difference applied across the conductor of length *l,* hence electric field produced at every point of the conductor is given by $E = V/l$

Hence electric force acting on every free electron, is

$$F = eE = \frac{eV}{l}$$

If m is the mass of electron, then acceleration experienced by the electron due to this force

$$a = \frac{F}{m} = \frac{eV}{ml} \qquad ...(2)$$

The electrons are not accelerated continuously but for a very short interval of time, called as Relaxtion time τ (= time taken between two successive collisions). Hence, velocity acquired by electron during relaxation time τ is given by

$v = 0 + a\tau = a\tau = (e\ v\ \tau/ml)$

After every collision with +ve ion, the velocity of electron becomes zero and again increases to v till it again collides with next +ve ion and so on. Hence average or drift velocity of electrons :

$$v_d = \frac{0+v}{2} = \frac{v}{2} = \frac{eV\tau}{2ml} \quad ...(3)$$

Substituting eq. (3) in eq. (1), we get

$$i = neA\frac{eV\tau}{2ml} = \left[\frac{ne^2\tau}{2m}\right]\frac{A}{l}V$$

or $$\frac{V}{i} = \left[\frac{2m}{ne^2\tau}\right]\frac{l}{A}$$

For a given metallic wire at a given temperature, all the quantities on R.H.S. of above equation will be constant. This mean that (V/i) = R = constant

This thus proves *ohm's law.*

RESISTIVITY OR SPECIFIC RESISTANCE

(i) We know that Resistance of a conductor of length l and C.S. area A is given by

$$R = \left[\frac{2m}{ne^2\tau}\right]\frac{l}{A}$$

Here quantities within bracket depend upon temperature and material of the conductor but those outside the bracket depend only on the dimensions of the conductor. Hence, if temperature and material of the conductor are kept constant, then quantity within the bracket will be a constant. This constant quantity is called as specific resistance and is represented by ρ.

(ii) Specific Resistance depends only on temperature and material of the conductor but not on its dimensions. As ρ depends only on the material of a conductor at a given temperature, hence it is a *characteristic constant.*

(iii) Hence $R = (\rho l/A)$ or $\rho = (RA/l)$;

If A = 1 and l = 1, then ρ = R

i.e. the specific resistance of the material of a conductor is equal to the Resistance of a conductor of that material, having unit length and unit cross sectional area, at a given temperature.

(iv) Alternatively, the specific resistance of a material is also defined as the resistance offered by an Unit Cube of that material between its two opposite faces, when current enters ⊥ to one face of cube and leaves ⊥ to opposite face.

(v) Unit of ρ in SI system = ohms x meter

(vi) $[\rho] = \frac{[R][A]}{[l]} = \frac{[V][A]}{[I][l]} = \frac{[W][A]}{[q][I][l]}$

$= \frac{[ML^2T^{-2}][L^2]}{[AT][A][L]} = [ML^3T^{-3}A^{-2}]$

(vii) **Relation between ρ, E and J**

$\rho = R\frac{A}{l} = \frac{V}{I} \times \frac{A}{l} = \frac{V/l}{I/A} = \frac{E}{J}$

(viii) **Specific conductance**

(a) Reciprocal of specific resistance is called as specific conductance and is represented by σ

(b) $\sigma = \frac{1}{\rho} = \frac{J}{E}$

or $J = \sigma E$

(c) Unit of s is ohm^{-1} $meter^{-1}$

(xi) The property of specific resistance of the materials can be used to select the proper material for making resistance wires needed for different purpose.

(a) *For making connection wires, we must use materials having low specific resistance* so that resistance of connection wires may be minimum possible and they may not effect the current in a circuit in which they are used.

As ρ for pure metals like silver, copper, gold, aluminium etc. is quite low, hence these materials are used for making connection wires.

(b) *For making resistance wires used for making resistance boxes, we must use materials having high specific resistance* so that a given resistance may be obtained by using a smaller length of the wire.

As specific resistance of the alloys like Constantan, Nichrome, Mangnin etc. is quite high, hence these alloys are used for making resistance wires used in resistance boxes.

Note : These alloys have one more special property that their *temperature coefficient of resistance (α) is quite.* Because of this property, the resistance of a Resistance box is negligibly effected by change of temperatue.

Some Important Points Concerning Resistance

(i) We know that $R = \left[\frac{2m}{ne^2\tau}\right]\frac{l}{A}$

For a given material of the conductor at a given temperature – R ∞ l and R ∞ (1/A)

Hence Resistance of a long and thin wire will be greater than that of a short and thick wire.

(ii) If a wire of a given resistance R is stretched to make its length doubled then its resistance becomes four times of its previous value.

Because volume will remain constant in the process of stretching, hence $Al = A'2l$ i.e. $A' = A/2$ i.e. New resistance of the stretched wire

$$R' = \rho\frac{l'}{A'} = \rho\frac{2l}{A/2} = 4\rho\frac{l}{A} = 4R$$

(iii) If a wire of given resistance R is melted and a fresh wire of double radius is prepared, then resistance of new wire is reduced to 1/16 of the previous value.

Again, because volume will remain constant, hence $Al = 4\,Al'$ i.e. $l' = l/4$

i.e. resistance of fresh wire

$$R' = \rho\frac{l'}{A'} = \rho\left[\frac{l/4}{4A}\right] = \frac{1}{16}\left[\frac{\rho l}{A}\right] = \frac{R}{16}$$

RESISTANCE OF PURE METALS

(i) We know that $R = \left[\frac{2m}{ne^2\tau}\right]\frac{l}{A}$

For a given conductor, l, A and n are constant, hence

$R \propto (1/\tau)$

If γ represents the mean free path (Average distance covered between two successive collisions) of the electron and v_{rms}, the root - mean - square speed, then

$$\tau = \frac{\lambda}{v_{rms}}, \text{ Hence } R \propto \frac{v_{rms}}{\lambda}$$

(a) λ decreases with rise in temperature because the amplitude of vibrations of the +ve ions of the metal increases and they create more hindrance in the movement of electrons and ,

(b) v_{rms} increase because $v_{rms} \propto \sqrt{T}$. Therefore, *Resistance of the metallic wire increases with rise in temperature.* As $\rho \propto R$ and $\sigma \propto (1/\rho)$, hence resistivity increase and conductivity decrease with rise in temperature of the metallic wires.

(ii) If R_0 and R_1 represent the resistances of metallic wire at 0°C and t °C respectively then R_1 is given by the following formula:

$$R_1 = R_0 (1 + \alpha t) \qquad ...(1)$$

where α is called as the *temperature coefficient of resistance* of the material of the wire.

From equation (1) we can write

$$\alpha = \frac{R_t - R_0}{R_0 \times t} \text{ per } {}^\circ C$$

If R_0 = 1ohms, t = 1 ° C, Then $\alpha = R_t - R_0$ i.e. the temperture coefficient of resistance of the material of a wire is equal to increase in its resistance on raising its temperature by 1°C if its resistance at 0°C is assumed to be 1 ohm.

(iii) For most of the metals, α is nearly equal to (1/273) per °C, hence according to equation (1)

$$R_t = R_0\left[1 + \frac{t}{273}\right] = R_0\left[\frac{273 + t}{273}\right] = R_0\left[\frac{T}{273}\right]$$

where T is the absolute temperature. Hence $R_t \propto T$

i.e. *Resistance of pure metallic wires or the resistivity of metals is directly proportional to absolute temperature.*

RESISTANCE OF ALLOYS

(i) Resistance or resistivity of alloys also increases with increase in temperature but the increment in Resistance is found to be much smaller as compared to pure metals.

(ii) *α for alloys is +ve and very small*

(iii) There are some special alloys such as constant, mangnin etc. whose temterature coefficient of resistance is negligible i.e. whose resistivity is negligibly effected by rise in temperature. As mentioned earlier also, it is because of their high resistivity and low temperature coefficient of resistance that these materials are used for making resistance wires used in standard resistance boxes.

RESISTANCE OF SEMICONDUCTORS

(i) There are certain substances whose conductivity lies in between that of insulators and conductors, higher than that of insulators but lower than that of conductors. These are called as semiconductors, e.g. silicon, germanium, carbon etc.

(ii) The resistivity of semiconductors decreases with increases in temperature i.e. *α for semicondutors is –ve and high.*

(iii) Though at ordinary temperature the value of n (no. of free electrons per unit volume) for these materials is very small as compared to metals, but increases very rapidly with rise in temperature (this happens due to breaking of covalent bonds). Therefore, the resistance

$R = \dfrac{2ml}{ne^2 \tau A}$ goes on decreasing with increase in temperature.

Note : It is worth mentioning that here also τ decreases with rise in temperature but increase in the value of n is much higher as compared to decrease in τ i.e. net effect is that resistance decreases with rise in temperature.

RESISTANCE OF ELECTROLYTES

Resistance or resistivity of electrolytes also decreases with rise in temperature. This all happens due to the fact that viscosity of electrolytes decreases with rise in temperature, so that ions now get more freedom of motion i.e. Resistance decreases with rise in temperature.

RESISTANCE OF SUPERCONDUCTORS

There are also some such substances whose resistance shows a special behaviour with change in temperature, in low temperature region. As the temperature of these substances is decreased, their resistance first decreases in ordinary way, just like pure metals, but after a certain fixed very low temperature the resistance decreases extremely rapidly and tends to zero e.g. resistance of mercury decreases very rapidly and tends to zero at 4°K. This phenomenon is called as superconductivity.

Generation of Heat by Electric Energy

(i) When the ends of a metallic wire are connected with the terminals of a battery, free electrons present in the conductor starts moving with a drift velocity in a direction opposite to that of applied electric field and a current is established in the conductor. These electrons are not accelerated indefinitely but constantly collide with +ve ions present in the conductor and transfer their energy obtained from the battery, to these +ve ions.

Thus electrons go on taking energy from the battery and losing it continuously in undergoing collisions with +ve ions. As a result of this, root mean square velocity of ions is increased i.e. temperature of wire is increased i.e. electrical energy is dissipated in the form of heat energy.

(ii) Let a current of i ampere is allowed to flow in a metallic wire for t seconds when a potential differences V is established between two ends of wire. If the charge q flows through the wire in t seconds, then q it.

Hence work done by the battery in taking q coulomb of charge from one end of wire to other, is given by

$$W = Vq = Vit \text{ Joules} \quad ...(1)$$

If R represents the resistance of metallic wire, then $V = iR$ or $i = (V/R)$

$$\text{Hence } W = i^2 Rt = (V^2t/R) \text{ Joules} \quad ...(2)$$

$$\text{Thus, } W = Vit = i^2 Rt = (V^2t/R) \quad ...(3)$$

Because 1 calorie of heat is equivalent to 4.2 Joules of work, hence heat produced or energy dissipated in the metallic wire in t sec is given by

$$H = \frac{W}{4.2} = \frac{Vit}{4.2} = \frac{i^2 Rt}{4.2} = \frac{V^2 t}{4.2 R}$$

INTERNAL RESISTANCE OF A CELL

When the terminals of a cell are connected by a wire, an electric current flows in the wire from positive terminal of the cell towards the negative terminal, but in side the electrolyte of the cell it flows from the negative terminal to positive terminal. Just as, we have read, material of the wire opposes the flow of current, the electrolyte of the cell also creates opposition or resistance in the path of current. This resistance, due to the electrolyte, is known as Internal resistance of the cell. Due to this resistance , a part of the energy supplied by the cell is dissipated in the cell itself in the form of heat energy.

Therefore work done by the cell or energy given by the cell for flow of charge q, is

$W = Eq = Eit.$

This energy is used up in two ways.

(a) *In external circuit :* If V is the potential difference across the external resistance R (which is also the potential difference between terminals of cell, when current is being drawn from the cell) then energy consumed in it (or work done outside the cell) is

$$W_{ext} = Vit$$

(b) *In internal circuit :* If V' is the potential drop in the electrolyte of the cell due to its internal resistance r, then energy consumed inside the cell (or work done inside the cell) is

$W_{int} = V' i t = i^2 r t$

Hence, according to law of conservation of energy

$W = W_{ext} + W_{int}$

$Eit = Vit + i^2 r t$

$E = V + ir$

or $V = E - ir$ ...(1)

Important Points

(i) If a cell is being discharge or current is being drawn from the cell, then potential difference across the terminals of cell is always less than its E.M.F. It happens due to a potential drop across the internal resistance of the cell.

(ii) Higher is the value of current drawn from the cell, lesser will be the potential difference across the terminals of the cell.

(iii) EMF is the characteristic property of the cell which remains constant for a cell but potential difference goes on decreasing as more & more current is drawn from the cell, due to increase in potential drop across the internal resistance.

(iv) If the cell is in open circuit or no current is being drawn from the cell, then from eq. (1) we get V = E i.e. potential difference between the terminals of cell is equal to E.M.F. of the cell.

Note : (a)Generally, a voltmeter is used to measure the E.M.F. of a cell. But when a voltmeter is connected across the terminals of cell, cell no more remains in open circuit because some current is drawn by the voltmeter. As a result of this, reading of voltmeter < E.M.F. of the cell.

(b) However, because of the resistance of voltmeter being quite high, current drawn by voltmeter is very small and therefore reading of voltmeter $\approx$ E.M.F. of the cell.

(c) E.M.F. can be measured accurately with the help of a potentiometer.

(v) If the cell is short circuited, then a maximum current is drawn from the cell (= E/r) and potential difference across its terminals becomes zero.

$$(V = E - ir = E - (E/r) \times r = 0)$$

(vi) If V is the potential difference across the external resistance R due to flow of current i then V = i R and according to eq. (1)

$$iR = E - ir$$

or $$E = i\,(R + r)$$

or $$i = \frac{E}{R + T} \qquad \text{...(2)}$$

If there are more than one cells and external resistances in the circuit, then equation (2) may be written in following generalised form

$$\text{Main current in a circuit} = \frac{\text{Net E.M.F.}}{\text{Total External Resistance} + \text{Total Internal Resistance}} \quad ...(3)$$

(vii) From eq (2) we can write :

$$\frac{V}{R} = i = \frac{E}{R+r}$$

or $$\frac{E}{V} = \frac{R+r}{R} = 1 + \frac{r}{R}$$

or $$r = R\left(\frac{E}{V} - 1\right) \quad ...(4)$$

Thus, r can be measured just by measuring the potential difference across the terminals of cell in two cases.

(a) when key K is open, then reading of voltmeter $\cong$ E.

(b) when key K is closed, then reading of voltmeter $\cong$ V.

KIRCHOFF'S LAWS

Kirchoff gave following two laws for determining the distribution of currents in different branches of a complicated electrical circuit.

First Law : *In a network of conductors, the algebric sum of currents meeting at any junction in the circuit is always equal to zero i.e.*

$$\Sigma i = 0.$$

While applying this law we have to follow a convention for deciding the sign of current. The currents approaching the junction are taken as positive while those leaving the junction are taken as negative. For example, in the figure, currents i_1 i_2 i_6 and i_7 are to be taken as positive while i_3 i_4 and i_5 as negative. Hence according to Kirchoff's 1st law

$$i_1 + i_2 + i_6 + i_7 - i_3 - i_4 - i_5 = 0$$

or $$i_1 + i_2 + i_6 + i_7 = i_3 + i_4 + i_5.$$

Thus, sum of currents approaching the junction is equal to the sum of currents leaving the junction or total charge approaching the junction per second is equal to that leaving the junction per second i.e. when a steady current flows in the circuit, then there is neither any collection of charge at any point in the circuit nor any charge is removed from there. In other words, *Kirchoff's first law is an alternative form of the law of conservation of charge.*

Second law : *In any closed mesh of a circuit, the algebric sum of the products of the currents and resistances of the different parts of the mesh is always equal to the algebric sum of different emf's acting in that mesh*

i.e. $\Sigma\, i\, R = \Sigma\, E$

Applying Kirchoff's 2nd law to meshes 1 and 2, we get

$$i_1r_1 - i_2r_2 = E_1 - E_2 \quad ...(1)$$

and $$i_2r_2 + (i_1 + i_2)\, R = E_1 \quad ...(2)$$

Solving these equations, we can find currents in different branches of the circuit.

COMBINATIONS OF RESISTANCES

A number of resistances can be connected in a circuit and any complicated combination can be, in general, reduced essentially to two different types, namely series and parallel combinations.

(i) In this combination current flowing through each resistance will be sáme and will be equal to current supplied by the battery.

(ii) As resistances are different and current flowing through them is same, hence potential differences across them will be different. Applied potential difference will be distributed among three resistances directly in their ratio.

As i is constant, hence V ∞ R

i.e. $V_1 = iR_1$, $V_2 = iR_2$, $V_3 = iR_3$.

(iii) If the potential difference between the points A and D is V, then $V = V_1 + V_2 + V_3 = i\,(R_1 + R_2 + R_3)$

(iv) If the combination of resistances between two points is replaced by a single resistance R such that there is no change in the current of the circuit and in the potential difference between those two points, then the single resistance R will be equivalent to combination and V = i R i.e.

$iR = i\,(R_1 + R_2 + R_3)$

or $\mathbf{R = R_1 + R_2 + R_3}$.

(v) Thus in series combination of resistances, important conclusions are :

(a) Equivalent Resistance > highest individual resistance.

(b) Current supplied by source = Current in each resistance

or $\dfrac{V}{R_1+R_2+R_3}=\dfrac{V_1}{R_1}=\dfrac{V_2}{R_2}=\dfrac{V_3}{R_3}$

(c) The total potential difference V between points A and B is shared among the three resistances directly in their ratio.

$V_1 : V_2 : V_3 = R_1 : R_2 : R_3$.

Resistances in Parallel

(i) When two or more resistances are combined in such a way that their first ends are connected to one terminal of the battery while other ends are connected to other terminal, then they are said to be connected in parallel. Three resistances R_1 , R_2 and R_3 joined in parallel between two points A and B. Suppose the current flowing from the battery is i. This current gets divided into three parts at the junction A. Let the currents in three resistances R_1 , R_2 and R_3 are i_1 , i_2 i_3 respectively.

(ii) Suppose potential difference between points A and B is V. Because each resistance is connected between same two points A and B, hence potential difference across each resistance will be same and will be equal to applied potential difference V.

(iii) Since potential difference across each resistance is same, hence current approaching the junction A is divided among three resistances reciprocally in their ratio.

As V is constant, hence i ∞ (1/R) i.e.

$$i_1=\frac{V}{R_1},\ i_2=\frac{V}{R_2}\text{ and } i_3=\frac{V}{R_3}$$

(iv) Because i is the main current which is divided into three parts 1_1 , i_2 and i_3 at the junction A, hence

$$i=i_1+i_2+i_3=V\left[\frac{1}{R_1}+\frac{1}{R_2}+\frac{1}{R_3}\right]$$

If the equivalent resistance between the points A and B is R, then i = V/R

Thus, $\dfrac{V}{R}=V\left[\dfrac{1}{R_1}+\dfrac{1}{R_2}+\dfrac{1}{R_3}\right]$

$$\text{or } \frac{1}{R} = \frac{1}{R_1} + \frac{1}{R_2} + \frac{1}{R_3}$$

(v) Thus in parallel combination of resistances important conclusions are :

(a) Equivalent Resistance < lowest individual resistance

(b) Applied potential difference = Potential difference across each resistance.
or $iR = i_1R_1 = i_2R_2 = i_3R_3$.

(c) Current approaching the junction A = Current leaving the junction B and current is shared among the three resistances in the inverse ratio of resistances

$$i1 : i2 : i3 = \frac{1}{R_1} : \frac{1}{R_2} : \frac{1}{R_3}$$

WHEATSTONE'S BRIDGE

(i) Wheatstone designed a network of four resistances with the help of which the resistance of a given conductor can be measured. Such a network of resistances is known as Wheatstone's bridge.

(ii) When key K_1 is pressed, a current i flows from the cell. On reaching the junction A, the current i gets divided into two parts i_1 and i_2. Current i_1 flows in the arm AB while i_2 in arm AD. Current i_1, on reaching the junction B gets further divided into two parts $(i_1 - i_g)$ and i_1, along branches BC and BD respectively. At junction D, currents i_2 and i_g are added to give a current $(i_2 + i_g)$ in the branch DC. Finally, currents $(i_1 - i_g)$ and $(i_2 + i_g)$ add up at junction C to give a current $(i_1 + i_2)$ or i along branch CE. In this way, currents are distributed in the different branches of bridge. In this position, we get a deflection in the galvanometer.

(iii) Now the resistances P, Q, R and S are so adjusted that on pressing the key K_2, deflection in the galvanometer becomes zero or current i_g in the branch BD becomes zero. In this situation, the bridge is said to be balanced.

(iv) In this balanced position of bridge, same current i_1 flows in arms AB and BC and similarly same current i_2 in arms AD and DC. In other words, resistances P and Q and similarly R and S, will now be joined in series.

(v) *Condition of Balance :* Applying Kirchoff's 2nd law to mesh ABDA,

$$i_1P + i_gG - i_2R = 0 \quad ...(1)$$

Similarly, for the closed mesh BCDB, we get

$$(i_1 - i_g) Q - (i_2 + i_g) S - i_gG = 0 \quad ...(2)$$

When bridge is balanced, $i_g = 0$.

Hence eq. (1) & (2) reduce to

$$i_1 P - i_2 R = 0 \text{ or } i_1 P = i_2 R \quad ...(3)$$

$$i_1 Q - i_2 S = 0 \text{ or } i_1 Q = i_2 S \quad ...(4)$$

Dividing (3) by (4), we have

$$\frac{P}{Q} = \frac{R}{S} \quad ...(5)$$

This is called as condition of balance for wheatstone's Bridge.

(vi) It is clear from above equation that if ratio of the resistances P and Q, and the resistance R are known, then unknown resistance S can be determined. This is the reason that arms P and Q are called as Ratio arms, arm AD as known arm and arm CD as unknown arm.

(vii) When the bridge is balanced then on interchanging the positions of the galanometer and the cell there is no effect on the balance of the bridge. Hence, the arms BD and AC are called as conjugate arms of the bridge.

(viii) The sensitivity of the bridge depends upon the value of the resistances. *The sensitivity of bridge is maximum when all the four resistances are of the same order.*

Potentiometer

(i) *It is an accurate device for measuring the emf of a cell or the potential difference between two points of an electric circuit.*

(ii) It consists of a very long (4 - 12 meters) and uniform wire which is made of a material having high specific resistance and low temperature coefficient of resistance such as constantan, nichrome etc. The wire is spread on a wooden board in the form of parallel wires, each of 1 meter length.

(iii) A cell C, whose e.m.f. is to be measured, is also connected with the wire in such a way that its +ve terminal is connected with one end P of the wire and –ve terminal is connected through a galvanometer G to a jockey J which is free to slide along the wire and can be made to touch it at any point.

(iv) When current from the battery flows through the wire from end P to end Q, a potential difference is produced across the wire. The fall of potential per unit length of the wire is called as Potential Gradient (= K).

(v) (a) *When jockey is pressed on some point L on extreme LHS,* then potential difference between point Pand L < e.m.f. of cell (= E_c).

Since potential at point P > potential at L, hence current from battery B (shown by → and represented by i_b) will flow through galvanometer along the path PCL. But because the +ve terminal of the cell C is also connected to the point P, hence current from cell (shown by ⇒ and represented by i_c) will flow through the galanometer along the path LCP i.e. two currents i_b, and i_c will be in opposite directions. But because e.m.f. of the cell is greater than potential difference between point P and L, hence $i_c > i_b$ i.e. resultant current will flow through the galvanometer in the direction LCP and needle of the galvanometer is deflected towards one side.

(b) *When jockey is pressed at some point R on extreme RHS,* then

Potential difference between point P and R > e.m.f. of cell E_c. and $i_b > i_c$. In this situation, resultant current will flow through the galvanometer in the direction PLC and the needle of the galvanometer is deflected towards opposite side.

(c) *When jockey is pressed at some point N in between the points L and R :* It is now evident that in between points L and R there will be a point, say N, such that when jockey is made to touch it, there will be no defection in galvanometer. This point N is called as Null point. In this situation. *Potential difference between point A and N will be equal to the e.m.f. of the cell.*

Note : Because in Null position, there is no current in the cell circuit, hence cell will be effectively in an open circuit and potential difference across its terminals will be equal to its e.m.f.

(vi) If i is the current flowing through the wire then in null position then, $E_c = V_{PN} = ir_{PN} = (i\,\rho\,l)/A$

where r_{PN} = resistance of the portion PN of the wire

l = length of portion PN or balance length

ρ = Resistivity of the material of the wire

A = Cross sectional area of the wire

Hence $(E_c/l) = (\rho/A)$ ...(1)

If R and L represent the resistance and length of the complete potentiometer wire then

$$R = \rho\left(\frac{L}{A}\right)$$

or $\dfrac{\rho}{A} = \dfrac{R}{L}$ = Resistance per unit length

If cross sectional area of the wire is uniform, then its resistance per unit length will be constant at constant temperature.

$$\text{Hence } \frac{E_c}{l} = \frac{\text{Current} \times \text{Resistance of wire}}{\text{length of wire}}$$

$$= \frac{\text{Potential difference across the wire}}{\text{Length of the wire}}$$

= Potential gradient = K i.e. $(E_c = K\,l)$

(viii) **Important Points**

(a) Higher is the length of potentiometer wire, lesser is the potential gradient along the wire and therefore higher is the balance length even for a very small e.m.f. which can be measured more accurately.

(b) The wire must be uniform otherwise K will be different at different points of the wire and readings will be inaccurate.

(c) E_b must be greater than E_c otherwise e.m.f. will not be balanced even over the complete length of wire.

(d) +ve terminals of both the battery and cell must be connected at same point P otherwise i_b and i_c will be in same direction and null point is never obtained.

Ohm's Law and Resistance

1. $I \propto V$ or $V = IR$ or $I = GV$, provided physical conditions (temperature, magnetic field, radiation etc.) remain unchanged. Where G = conductance = $\frac{1}{R}$. [R = resistance].

   $G = \frac{ne^2\tau}{2m}\left(\frac{A}{l}\right)$, $G = J/E$. Where m = Mass of electron, τ = Relaxation time, E = Intensity of electric field, n = Number of free electrons A = Area, l = length.

   S.I. unit of G = mho = $ohm^{-1} = \frac{1}{ohm} = \frac{amp}{volt}$ siemen.

2. $R = \rho\frac{l}{A}$. where l = length of wire (along the direction of flow of I), A = cross sectional area ($\perp$ to the direction of flow of I) & ρ = specific resistance or electrical resistivity.

   σ (electrical conductivity or specific conductance) = $\frac{1}{\rho}$. Specific conductance $(\sigma) = \frac{J}{E}$

   Where J = Current density, E = Intensity of electric field.

   S.I. unit of ρ & σ = ohm (Ω) × metre & mho/metre respectively.

3. $R = \rho\frac{l}{\pi r^2}$. ($A = \pi r^2$ for wire) $= \rho\frac{l/\pi D^2}{4}$. Where D = diameter.

4. $R = \frac{\rho V}{\pi^2 r^4}$. Where V is the volume.

5. $R_t = R_0(1 + \alpha t)$. Where α = temperature co-efficient of R. Unit of α is per ^0C.

6. *To be remembered :*

   (i) 1 amp = 10^{-1} e.m.u. of I = 3×10^9 e.s.u. of current.

   (ii) 1 coul = 10^{-1} e.m.u. of charge = 3×10^9 e.s.u. of charge.

   (iii) 1 volt = 10^8 e.m.u. of potential difference or 10^8 ab bolt $= \frac{1}{300}$ e.s.u. of potential difference or $\frac{1}{300}$ stat volt.

   (iv) 1 ohm = 10^9 e.m.u. of R = $\frac{1}{300}$ stat volt.

7. *Series grouping* : $R_s = r_1 + r_2 + r_3 + ... + r_n$.

   The equivalent conductance is given by

$$\frac{1}{G} = \frac{1}{g_1} + \frac{1}{g_2} + \frac{1}{g_3} + ... \frac{1}{g_n}$$

$$V = V_1 + V_2 + V_3 + ... V_n.$$

   In a series potential difference across any resistance (r) is

$$v' = \frac{r'}{r_{eq}} V$$

   where r_{eq} = equivalent resistance of series

   V = voltage across series, for Ex. – from figure

$$V_1 = \frac{R_1}{R_1 + R_2 + R_3} . V .$$

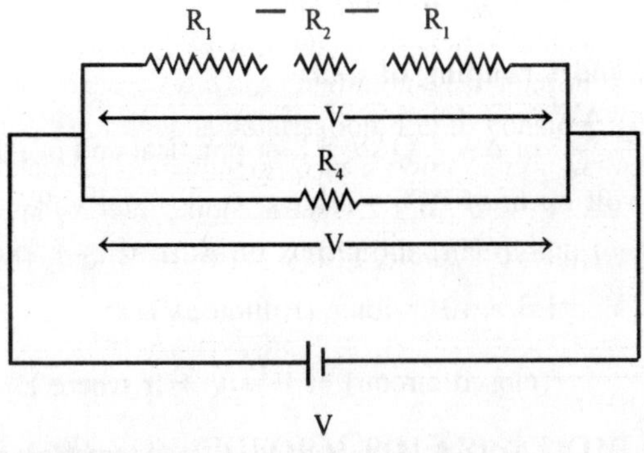

Fig. 1

8. *Parallel grouping* :

$$\frac{1}{R_p} = \frac{1}{r_1} + \frac{1}{r_2} + \frac{1}{r_3} + ... + \frac{1}{r_n} .$$

or $$G = G_1 + G_2 + G_3 + ... + G_n.$$

$$I = I_1 + I_2 + I_3 + ... I_n.$$

   In parallel, current through any Branch

$$i_1 = i \times \frac{R_2}{R_1 + R_2} .$$

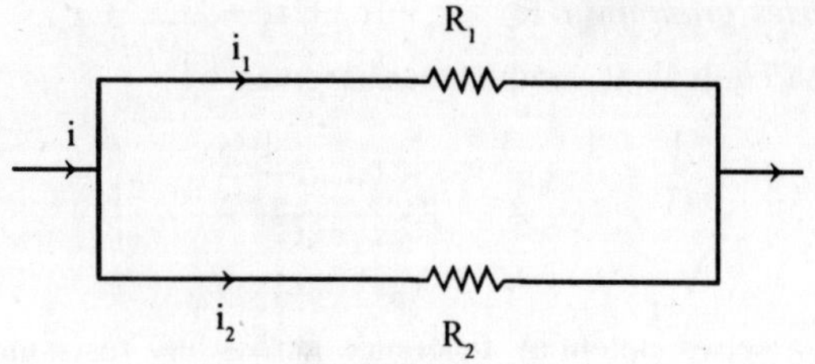

Fig. 2

9. *For two resistances in parallel :* $R_p = \frac{r_1 r_2}{r_1 + r_2} = \frac{\text{product}}{\text{sum}}$.

10. *For n identical resistances :* R_s = nr (series).

    $R_{p'} = \frac{r}{n}$ (parallel) and $\frac{R_s}{R_p} = n^2$.

11. Approximate percentage change in R = 2x small percentage changes in length by stretching.

E.M.F., P.D. and Grouping of Cells

1. $\Delta V = \frac{\Delta W}{Q}$ or $\Delta W = Q\Delta V$. S.I. or practical unit of P.D. = Joule/coul = volt e.s.u. of P.D. = erg/stat coul = stat volt.

2. 1eV = 1.6×10^{-19} Joule. 1 MeV = 1.6×10^{-13} Joule (Million eV).

   1 BeV = 1.6×10^{-10} Joule (Billion eV).

3. $I = \frac{E}{R + r}$ (closed circuit) of E = V + Ir where E = e.m.f.

   V = P.D., r = internal resistance, R = external resistance and Ir = potential drop.

   If, R = 0 (short circuit), I will be maximum, $I_{max} = \frac{E}{r}$.

   If, R = $\propto$ (open circuit), I will be minimum, $I_{min} = 0$.

4. *Series grouping of cells :* $I = \frac{nE}{nr + R}$

   $I_{max} = \frac{E}{r}$, where r >> R.

5. *Parallel grouping of cells :* $I = \frac{nE}{r + nR}$; $I_{max} = \frac{nE}{r}$,

when r >> R.

6. *Mixed grouping of cells* : $I = \frac{mnE}{nr + mR}$; $I_{max} = \frac{nE}{2R} = \frac{mE}{2r}$, when nr = mR.

   Where n = number of cells in one row, m = number of rows and m × n = total number of cells.

7. *Wrong series connection* : $I = \frac{(n-2m)E}{R + nr}$;

$$I_{max} = \frac{(n-2m)E}{nr},$$

   when R << r. Where n = total number of cells and m = number of cells wrongly connected.

Wheatstone Bridge & Kirchoff's Law

1. When galvanometer shows no deflection, PR = QS (i.e. products of alternate arms resistances are equal) or $\frac{P}{Q} = \frac{S}{R}$.
2. **Krichoff's Laws**

   Ist law : $\Sigma I = 0$ 2nd law : $\Sigma I\ R = \Sigma E$.

   Where E = total e.m.f. of the mesh or circuit.

HEATING EFFECT OF CURRENT AND THERMO-ELECTRICITY

1. Q = It. Where Q = charge, I = current, t = time.
2. **P.D.** across ends of conductor, V = IR. V = Voltage, I = Current, R = Resistance.
3. *Work done (W)* = Q.V. = VIt = I^2 Rt = $\frac{V^2}{R} \times t$ = P × t.
4. *Power consumption (P)* = $\frac{W}{t} = I\ V = I^2\ R = \frac{V^2}{R}$.
5. *Heat produced (H)* = $\frac{W}{J} = \frac{VIt}{J} = \frac{I^2Rt}{J} = \frac{V^2}{R} \times \frac{t}{J} = \frac{P}{J} \times t$.
6. *Energy* = P × t.
7. *Practical unit* of energy supply = killo-watt-how (kWh). 1 I.B.O.T. unit = 1 kWh = 36 × 10^5 Joules. 1 H.P. = 746 wolt.
8. *Number of units consumed* = $\frac{\text{watt} \times \text{hr}}{1000}$ = kWh.

S.I. unit : I → amp, V → volt, R → ohm, t → sec, W → Joule, P → watt, H → Joule.

M.K.S. Unit : 1 cal = 4.2 Joule, H → cal, J = 4.2 Joule/cal.

9. Heat produced H α R if I and t are constant, i.e. $\frac{H_1}{H_2} = \frac{R_1}{R_2}$.
10. Heat produced $H \propto \frac{1}{R}$ if V and t constant, i.e. $\frac{H_1}{H_2} = \frac{R_2}{R_1}$.
11. $H \propto I^2$ if R and t are constant.
12. H α t if R and I are constant.
13. *Hot wire instrument* (for measuring both A.C. & D.C.) :

    $\theta \propto H$ or $\theta \propto i^2$, i.e. $\frac{\theta_1}{\theta_2} = \frac{i_1^2}{i_2^2}$.
14. $t_n = \frac{t_i + t_c}{2}$. Where t_n = neutral temperature t_i = inversion temperature & t_c = temperature of cold junction.

Chemical Effect of Current

1. **Faraday's first law :**

   By flowing a charge Q

   through a electrolytic if mass m of substance liberated or deposited at the electrode

   $m \propto Q \rightarrow m = zQ = Z$ if z = Electrochemical equivalent of substance

   Second law : current flowing through different electrolytes.

   Then

   $m_1 : m_2 : m_3 ..... = E_1 : E_2 : E_3$

   $\therefore \quad m \propto E$

   E = chemical equivalent of substances.
2. Charge on one ion = ne. Where n = valency of ion, e = electronic charge = -1.6×10^{-19} coulomb.
3. Number of ions liberated at electrode = $\frac{Q}{ne}$. Where Q = charge flowing in electrolyte.
4. Mass of one atom = $\frac{A}{N}$. Where A = atomic weight N = Avogadro's number.

5. Mass of element (w) liberated at electrode = number of ions liberated × mass of one atom.

   i.e. $w = \frac{Q}{ne} \times \frac{A}{N} = \frac{1}{Ne} \times \frac{A}{n} \times Q = \frac{E}{F} \times Q$. Where E = eq. wt.

6. $\frac{E}{F} = Z$ (electrochemical equivalent). S.I. unit of Z is kg/coul.

SHUNT, AMMETER, VOLTMETER AND BRANCHING OF CURRENT

1. To increase the range of an ammeter or to convert a galvanometer into an ammeter, a low resistance (shunt) is connected in parallel.

   $S = \frac{G}{n-1}$ (in parallel). $n = \frac{I}{I_g}$.

   Where S = shunt resistance,

   G = resistance of galvanometer

   $\frac{I}{I_g} = 1 + \frac{G}{S}$

   Where I_g = full scale defection current or current which the ammeter can measure.

   I = Current to be measured.

2. To increase the range of a *voltmeter* or to convert an *galvanometer* into a *voltmeter*, a high resistance (R) is connected in series :

   $R = G(n-1) = \frac{V}{Ig} - G$, where G = resistance of galvanometer.

   $n = \frac{V}{V_g}$ or $= \frac{\text{new range}}{\text{old range}}$. Where V_g or V_1 = voltage which the voltmeter can measure and V or V_2 = voltage to be measured. Where i_g = current flows through galvanometer, V = voltage to be measured.

3. Branching of current in two parallel resistances :

   $I_1 = \frac{Ir_2}{r_1 + r_2}$ and $I_2 = \frac{Ir_2}{r_1 + r_2}$.

4. Ater shunting the decrease in resistance $= \frac{G^2}{G+S}$.

Electric Charge and Current

1. $I = \frac{\Delta Q}{\Delta t}$.
2. Electric charge is a fundamental property.
3. $Q = \pm ne = It$ (I = current).
4. Units S.I.–coloumb

   1 coulomb = 6.25×10^{18} electron.
5. Strength of storage cell (A.H.) = charging current (amp) × time (hr). [1 amp × hr = 3600 coulomb].
6. Current density $\vec{j} = \frac{I}{A}$ Amp/m^2.
7. (A = Area of cross section perpendicular to the direction of (I)
8. v_d (drift velocity of electron) =

   $$\frac{I}{neA} = \frac{J}{ne} = \frac{\sigma E}{ne} = \frac{E}{\rho\, ne} = \frac{V}{\rho\, lne}$$

   E = Intensity of electric field

   V = P.D. across conductor.
9. If length of a wire increased by x% then its resistance increases by 2x%.
10. If length of wire becomes n times, It's resistance becomes n^2 times.

Electric Current

(i) Electric current arises due to continuous flow of charged particles or irons.

(ii) Any medium, which have practically free electric charges capable of moving from one place to other, is a conductor of electricity i.e., it can carry an electric current. e.g.

(a) A solid metal consists of-vely charged free electrons and +vely charged ions. Free electrons can migrate from one place to other but +vely charged ions are fixed in their mean positions and can only vibrate about their mean positions This is a solid state conductor, in which motion of electrons constitutes an electric current.

(b) An electrolyte consists of equally and oppositely charged ions (as a result of its dissociation), both of which are capable of moving in the medium. This is a solid state conductor, in which motion of electric current is constituted due to motion of both +ve and- ve ions.

(iii) During the flow of current in a medium, the electric charges always flow from higher potential energy state to lower potential energy state. Therefore, a +ve charge always moves from higher electric potential to lower electric potential while a-ve charge from a lower electric potential to higher electric potential.

(iv) Electric charge flowing per second in an electric circuit is a measure of electric current in that circuit. If a charge q flows through an electric circuit in time t seconds, then electric current in that circuit is

$$i = q/t$$

If a small charge dq flows in time dt, then electric current

$$i = dq/dt$$

(v) The direction of current is *conventionally* taken as the direction of flow of +ve charge. If in a medium, current is due to flow of -ve charges, then direction of current will be opposite to direction of motion of-ve charges.

(ii) The unit of electric current in SI system is Ampere.

$$1 \text{ Ampere} = \frac{1 \text{ coulomb}}{1 \text{ second}}$$

(vii) Because in metals, current is due to flow of electrons and charge on one electron is equal to 1.6×10^{-19} coulomb, hence for one coulomb of charge $[(1/1.6 \times 10^{-19}) = (6.25 \times 10^{18})]$ electrons must flow i.e.

1 Ampere = 6.25×10^{18} electrons/sec.

This means that in one second 6.25×10^{18} electrons enter at one end of wire and same number of other electrons leave at the other end.

CURRENT DENSITY

(i) The current density at any point inside a conductor is defined as a vector quantity, having magnitude equal to current through a very small unit area around that point, the area being normal to direction of flow of current and direction along the direction of current at that point.

(ii) If A_n represent a small area at the point P, normal to direction of current 1, then

Current density $J = \frac{I}{A_n}$ ...(1)

If the plane of small area is not normal to the current but makes an angle θ with the area normal to current, then,

$$J = \frac{I}{A_n} = \frac{I}{A \cos \theta} \quad ...(2)$$

(iii) Unit of current density in SI system is ampere/meter^2.

(iv) Because current density and area are vector quantities, hence equation (2) can also be written as

$$I = JA \cos \theta = \vec{J} . \vec{A} \quad ...(3)$$

Electric Power

(i) The rate at which electric energy is consumed in an electric circuit, is called as its Electric power and is represented by P.

(ii) Hence from (3) we can write :

$$P = Vi = i^2 R = (V^2/R) \quad ...(4)$$

(iii) The unit of power is *Watt.* If in an electric circuit, energy is dissipated at the rate of 1 Joule/sec, then the power of the circuit is 1 Watt i.e. 1 Watt = (1 Joule/1 sec)

We can also write from eq. (4) –

1 Watt = 1 Volt × 1 Ampere

(iv) 1 KW = 1 Kilo Watt = 10^3 Watt,

1 MW = 1 Megha Watt = 10^6 Watt

1 mW = 1 milli Watt = 10^{-3},

Also, 1 Horse power = 746 Watt.

(v) It is also clear from eq. (4), *that at constant temperature, rate of production of heat in a current-carrying wire is either directly proportional to the square of the current or to the square of the potential difference.* This is known as Joule's law.

Note : It is to be noted here that if ohm's law remains valid, Joule's law also remains valid.

Some important conclusions from the concept of Electric Power

(i) If 100W-220 V is written on some electric instrument, then it simply means that on using the instrument on 220 volt supply, 100 Joules of energy is consumed by it one second.

(ii) *Other informations from voltage and wattage* : Let us take an example of a bulb whose wattage is P watts and which is manufactured for working on a supply of V volts (which is constant and equal to 220 volts). From this data we can find out the following-

(a) *Resistance of the filament of bulb*

$R = V^2/P$

As V is constant, hence *higher is the Wattage of a Bulb, lesser is the resistance of its filament or thicker is the filament* ($R \propto 1/A$).

(b) *Maximum current that can be allowed to pass through a bulb-*

$i_{max} = P/V$

i.e. higher is the wattage of bulb, higher is current that can be allowed to pass through a bulb.

Note : Value of i_{max} is fixed for a given bulb. If a bulb is being used in a circuit and current in that circuit is higher than i_{max} for that bulb, then bulb get fused.

(iii) If bulbs of different wattage are joined in parallel, then highest wattage bulb glows with maximum brightness.

Reason : As bulbs are joined in parallel, hence voltage across each bulb will be same. Now, because resistance of highest Wattage bulb is minimum, hence *heat produced (= V^2 t/4.2R) (or brightness) will be maximum in highest wattage bulb.*

This is the reason that connections in houses are made in parallel.

Note : There is one more advantage of making connections in parallel. If any branch of the circuit goes out of order, then rest of the circuit will go on working. Moreover, there will be an independent on-off switch for every appliance being used in the circuit of a house.

(iv) If bulbs of different wattage are joined in series, then lowest wattage bulb glows with maximum brightness.

Reason : As bulbs are joined in series, hence current through each bulb will be same. Now because resistance of lowest wattage bulb is maximum, hence *heat produced* (= i^2 *Rt/4.2) will be maximum in lowest Wattage bulb.*

Kilowatt-hour

(i) It is a commerical unit used for expressing consumed electric energy.

(ii) 1 KWH or 1 unit is the quantity of electric energy consumed in one hour in an electric circuit having power of 1 Kilo Watt.

(iii) 1 KWH = 1000 watt × 1 hour = [(1000 Joules/second) × (3600 seconds)] = *36 × 10⁵ Joules*

(iv) *Number of units consumed in an electric circuit*

If a current of i ampere flows in a circuit for t hours under a potential difference of V volts, then electric energy dissipated in the circuit = Electric power × time

= Vi (Watt) × t (hours)

= [(Vi/1000) (Kilo Watt)] × t (hours)

= (Vit/1000) Kilo Watt hour

∴ number of units

$$= \frac{\text{Volt} \times \text{Ampere} \times \text{hours}}{1000} = \frac{\text{Watt} \times \text{hours}}{1000}$$

$$= \frac{(\text{Ampere})^2 \times \text{ohm} \times \text{hours}}{1000}$$

E.M.F. of a Cell

We know that when electrons move through a conductor, their electrical energy is always consumed in undergoing collisions with ions of the conductor. Hence in order to maintain the flow of charge or electric current in an electric circuit continuously, some external agency is required for providing energy to the electrons for flowing through the whole circuit. The cell plays the role of this external agency. The energy liberated in the cell due to chemical reactions taking place in it, maintains the flow of charge in the circuit. In this way, chemical energy of the cell gets converted into electrical energy which in turn is dissipated in the form of heat energy. *Whatever work is done by the cell (or energy is given by the cell) for flow of unit charge throughout the whole circuit (including the cell) is called as E.M.F. of the cell.*

If W is the work done by the cell for flow of charge q throughout the whole circuit, then E.M.F. of a cell E = W/q

POTENTIAL DIFFERENCE

Suppose a source of constant e.m.f. E is connected in a circuit consuming of three resistances R_1 R_2 & R_3 joined end to end. If W is the energy given by the cell for flow of charge q throughout the circuit, then E = W/q.

If W_1 , W_2 and W_3 are the energies consumed in different parts of the circuit due to flow of charge q, then $W = W_1 + W_2 + W_3$.

and $E = \frac{W}{q} = \frac{W_1 + W_2 + W_3}{q} = \frac{W_1}{q} + \frac{W_2}{q} + \frac{W_3}{q}$

or $E = V_1 + V_2 + V_3$

where $V_1 = \frac{W_1}{q}$, $V_2 = \frac{W_2}{q}$ and $V_3 = \frac{W_3}{q}$

clearly represent the energies consumed in different parts of external circuit due to flow of an unit charge. These are known as *Potential Differences* across different parts of external circuit. Potential difference is measured by voltmeter.

Note : The potential difference between the two terminals of a surce, when no energy is drawn from it,is called as its Electromotive Force or in other words emf of a source is equal to the potential difference between its terminals when no current is drawn from it or when the source or cell is in open circuit.

SOLVED EXAMPLES

Example 1:

A copper wire having cross sectional area of 0.5 mm^2 and length of 0.1 m is initially at 25° C and is thermally insulated from its surroundings. If a current of 10 amps is set up in the wire-

(i) Find the time in which the wire will start melting. The change of resistance with the temperature of the wire may be neglected.

(ii) What will be the time if length of wire is doubled ? Given for copper, density = 9×10^3 kg/m^3.

Specific heat = 9×10^{-2} Kcal/kg- °C, melting point = 1075°

C, specific resistance $= 1.6 \times 10^{-8}\ \Omega$ *-m.*

Solution :

(i) If l is the length of the wire, A is the cross sectional area and ρ is the specific resistance of the material then resistance R of the wire is given by $R = \rho l/A$

If i is the current flowing in the wire, then electric energy supplied in t seconds, for increasing the temperature of the wire upto melting point, is given by

$$H = \frac{i^2 Rt}{4.2} = \frac{i^2 \rho l t}{A \times 4.2} \text{ cal} \qquad ...(1)$$

Mass of the wire = m = l Ad

Heat energy required for increasing the temperature upto melting point, is given by

$$H = m\ s\ (\theta_2 - \theta_1) = l\ A\ d\ s\ (\theta_2 - \theta_1) \qquad ...(2)$$

Hence from equations (1) & (2)

$$\frac{i^2 \rho l t}{A \times 4.2} = l\ A\ d\ s\ (\theta_2 - \theta_1)\ ;$$

$$t = \frac{A^2 ds\ (\theta_2 - \theta_1) \times 4.2}{i^2 \rho} \qquad ...(3)$$

$$\text{or}\quad t = \frac{(5 \times 10^{-7})^2 \times 9 \times 10^3 \times 9 \times 10^1 \times 4.2 \times 1050}{(10)^2 \times 1.6 \times 10^{-8}}$$

$$= 558.14 \text{ second.}$$

(ii) Because t is independent of the length of the wire, hence there will be no change in time if length of wire is doubled.

Example 2(a):

A heater using a Nichrome element operates at 220 volts. When it is switched on at 0° C, it carries an initial current of 2.75 ampere. After few seconds, the current reaches a steady value of 2.44 amp. What is the final temperature of the element ? The average value of the temperature coefficient of the resistance of Nichrome over the temperature range is 0.0045/° C.

Solution :

Given that at 0° C, V = 220 volts, i = 2.75 amp

Hence resistance of the wire at 0° C

R_0 = (V/i) = (220 / 2.75) = 80 ohms

If final temperature is t°C, then resistance of the wire at t°C

R_t = (220/2.44) = 90.16 ohms;

Now, $R_t = R_0 (1 + \alpha t)$

$$\therefore \quad t = \frac{R_1 - R_0}{R_0 \times \alpha} = \frac{90.16 - 80}{80 \times 45 \times 10^{-5}} = \mathbf{282^\circ C}$$

Example 2(b):

The density of platinum is 21.4 gm/cm³ and its resistivity is 1.0 × 10⁻⁷ ohm-m. Assume the wire to be rectangular with dimension of 0.01 cm and 0.02 cm. If the mass of the wire is 51.36 mg, what is the resistance of the wire ?

Solution :

If L is the length and A the cross sectional area of the wire, then

$$L\,A = \text{Volume of wire} = \frac{\text{mass}}{\text{density}} = \frac{m}{d},$$

$$\therefore \quad L = \frac{m}{Ad}$$

Hence Resistance of the wire is given by

$$R = \rho\frac{L}{A} = \rho\frac{m}{A^2 d}$$

$$= \frac{10^{-7} \times 51.36 \times 10^{-6}}{(2 \times 10^{-8})^2 \times 21.4 \times 10^3} = \mathbf{0.6\ ohms}$$

Example 3:

A 200 V DC generator transmits power to a distant place through cables of total length 5 km. If the cable has resistance of 0.08 ohm / km, what is the efficiency of transmission when the cable current is 150 A ? How is the efficiency altered by transmitting the same power at 500V?

Solution :

(i) Potential difference across the cable

= current × resistance of the cable

= 150 amp × (.08 ohms/km) × 5 km = 60 volt

Output potential = 200 – 60 = 140 volt

∴ Efficiency of transmission

$$= \frac{\text{Output voltage}}{\text{Input voltage}} \times 100 = \frac{140 \text{ Volt}}{200 \text{ Volt}} \times 100$$

= 70%

(ii) Delivered power = Vi = 200 × 150 = 30,000 Watts, Generated emf = 500 volt

It i is the current transmitted, then Voltage drop across the cable

= i × 0.4 = 0.4 i Volt

Since transmitted power is same in both cases, the current

$$= \frac{200 \times 150}{500} = 60 \text{ amp}$$

∴ Voltage drop across cable

= 0.4 × 60 = 24 volt

$$\text{Hence efficiency} = = \frac{500-24}{500} \times 100$$

$$= \frac{476 \times 100}{500} = \mathbf{95.2\%}$$

Example (4a):

Two metallic conductors, one made of copper and other of iron are connected in parallel across some source of constant e.m.f. At 20° C, currents flowing in both are equal. If temperature is raised to 100° C, what proportion of current will pass through each conductor. Given that α for Cu = 0.0042/° C and α for Iron = 0.006/° C.

Solution :

Because currents flowing in both the conductors are equal at 20° C, hence their resistance are equal at 20° C.

If R is the resistance of each conductor at 20°C, then resistance of copper conductor at 100° C.

$R_1 = R\,[1 + \alpha_1 \times 80] = R\,[1 + .0042 \times 80]$

= 1.336 R and that of iron conductor at 100°C

$R_2 = R[1 + \alpha_2 \times 80] = R[1 + .006 \times 80]$

$= 1.48\ R$

Because R_1 and R_2 are joined in parallel hence currents through them will be inversely proportional to their resistance i.e.

$$\frac{i_1}{i_2} = \frac{R_2}{R_1} = \frac{1.48\ R}{1.336\ R} = \frac{1480}{1336}$$

or, $$\frac{i_1 + i_2}{i_2} = \frac{1480 + 1336}{1336} = \frac{2816}{1336},$$

$$\therefore \quad \frac{i_2}{i_1 + i_2} = \frac{1336}{2816}$$

i.e. percentage proportion of i_2

$= (1336/2816) \times 100 = 47.44\%$.

$\therefore$ percentage proportion of $i_1 = 52.56\%$

Example 4(b):

Two heaters A and B are connected in parallel across the suply voltage V. Heater A generates 500 × 10^3 calories of heat in 20 minutes while B generates 1000 × 10^3 calories in10 minutes. The resistance of heater A is 10 ohms. What will be the resistance of heater B ? If these heaters are connected in series across the same voltage, how much heat will be produced in 5 minutes ?

Solution :

Because two heaters are connected in parallel across the suply voltage V hence voltage across each = V. Hence heat produced by two heaters are given by

$$H_1 = \frac{V^2 t_1}{4.2 R_1} \text{ and } H_2 = \frac{V^2 t_2}{4.2 R_2} \quad \text{...(1)}$$

$$\therefore \quad \frac{H_1}{H_2} = \frac{R_2}{R_1} \times \frac{t_1}{t_2}$$

or $$\frac{500 \times 10^3}{1000 \times 10^3} = \frac{R_2}{R_1} \times \frac{20}{10}$$

As $R_1 = 10$ ohms, hence $R_2 = (R_1/4) = 2.5$ ohms

when two heaters are connected in series, their total resistance

$= R_1 + R_2 = 12.5$ ohms

$\therefore$ Heat produced by combination

$$H = \frac{V^2}{4.2\,R}\,t \text{ calories} = \frac{V^2 \times 300}{4.2 \times 12.5} \text{ calories} \quad ...(2)$$

From equation (1),

$$H_1 = \frac{V^2\,t_1}{4.2\,R_1} \quad ...(3)$$

Dividing eq. (2) by eq. (3),

$$\frac{H}{H_1} = \frac{300}{t_1} \times \frac{R_1}{12.5}$$

$$\therefore \quad H = \frac{300}{t_1} \times \frac{R_1}{12.5} \times H_1$$

$$= \frac{300 \times 10 \times 500 \times 10^3}{20 \times 60 \times 12.5} = \mathbf{100 \times 10^3 \text{ calories}}$$

Example 4(c):

A wire of length 1.0 m and radius 10^{-3} m is carrying a heavy current and is a assumed to radiate as a black body. At equilibrium its temperature is 900 K white that of the surroundings is 300 K. The resistivity of the material of the wire at 300 K is $\pi^2 \times 10^{-8}$ Ω m and its temperature coefficient of resistance is 7.8×10^{-3} per ^{o}C. Find the current in the wire. (Given : Stefan Constant = 5.68×10^{-8} w m^{-2} K^{-4})

Solution :

Given that length of the wire l = 1.0 m, radius of the wire r = 10^{-3} m, equilibrium temperature $T_2 = 900$°K, Temperature of surroundings T_1 = 300°K, ρ_0 = Resistivity at 300°K = $\pi^2 \times 10^{-8}$ Ω m,

$$\alpha = 7.8 \times 10^{-3}/°C$$

and $\sigma = 5.68 \times 10^{-8}$ $Wm^{-2}K^{-4}$.

Let A = Curved surface area of the wire through which energy is radiated out = $2\pi r l$.

A' = Cross sectional area of the wire = πr^2.

In equilibrium,

$$\left.\begin{array}{l}\text{Energy lost/sec, due to radiations}\\\text{emitted by the wire}\end{array}\right] = \left[\begin{array}{l}\text{Energy produced per sec due}\\\text{to flow of Current}\end{array}\right.$$

According to Stefan's Law, energy lost per sec through curred surface area

$$= \sigma A (T_2^4 - T_1^4) = \sigma 2\pi r l (T_2^4 - T_1^4)$$

According to Joule's law of heating, energy produced per sec due to flow of current

$$= I^2 R = I^2 \rho \frac{l}{A'} = I^2 \rho \frac{l}{\pi r^2} = I^2 \rho_0 (1 + \alpha d\theta) \frac{l}{\pi r^2}$$

Hence, $\sigma\ 2\pi r l\ (T_2^4 - T_1^4) = I^2\ \rho_0\ (1 + \alpha d\theta)(l/\pi r^2)$

$$\text{or} \quad I^2 = \frac{\sigma 2\pi r l (T_2^4 - T_1^4)\pi r^2}{\rho_0 (1 + \alpha\, d\theta) l}$$

$$= \frac{\sigma \times 2\pi^2 r^3 (T_2^4 - T_1^4)}{\rho_0 (1 + \alpha\, d\theta)}$$

$$= \frac{(5.68 \times 10^{-8}) \times 2\pi^2 \times (10^{-3})^3 \times (900^4 - 300^4)}{(\pi^2 \times 10^{-8})[1 + (7.8 \times 10^{-3}) \times (900 - 300)]}$$

$\therefore$ I = 35.5 amp.

Example 4(d):

A charged capacitor of 5 × 10⁻² F capacity is discharged through a resistor R of 20 Ω and a copper voltmeter of internal resistance 30 Ω connected in series. If 4.62 × 10⁻⁶ kg copper is deposited, calculate the heat generated in the resistor R.(ECE of copper is 3.3×10⁻⁷ kg/C)

Solution :

If charge on the plates of capacitor is q, then from Faraday's law of electrolysis

$$m = Zq$$

$$\text{or} \quad q = \frac{m}{Z} = \frac{4.62 \times 10^{-6}}{3.3 \times 10^{-7}} = 14 \text{ Coul}$$

Hence energy stored in the capacitor

$$U = \frac{q^2}{2C} = \frac{(14)^2}{2 \times 5 \times 10^{-2}} = 1960 \text{ Joule}$$

In series the current i is same and heat energy produced, $H = i^2Rt$

or $H \infty R$

$\therefore$ Heat generated in the resistor R

$$H_1 = \frac{20}{20+30} \times \text{total Energy (U)} = \frac{20}{50} \times 1960$$

= 784 Joule.

Example 5:

A uniform copper wire of mass 2.23×10^{-3} kg carries a current of 1 A when 1.7 V is applied across it. Calculate its length and area of cross-section. If the wire is uniformly stretched to double its length, calculate the new resistance. Density of copper is 8.92×10^3 kg m^{-3} and resistivity is 1.7×10^{-8} Ωm.

Solution :

$$R = \text{Resistance of wire } \frac{V}{I} = \frac{1.7}{I} = 1.7 \text{ ohms}$$

Now, $R = \rho \dfrac{l}{A}$

$$\text{or } \frac{l}{A} = \frac{R}{\rho} = \frac{1.7 \text{ ohms}}{1.7 \times 10^{-8} \text{ ohms} \times \text{meter}}$$

= **10^8/meter**

$\therefore A = 10^{-8}\, l$ ...(1)

Volume of the wire

$$= lA = \frac{m}{d} = \frac{2.23 \times 10^{-3} \text{ kg}}{8.92 \times 10^3 \text{ kg/m}^3}$$

or $lA = 2.5 \times 10^{-7}$ m^3 ...(2)

From (1), and (2)

$$l^2 = \frac{2.5 \times 10^{-7}}{10^{-8}} = 25$$

or I = 5 m

From (1), $A = 10^{-8} \times 5 = 5 \times 10^{-8}$ m^2.

When the wire is stretched 2 times, volume remains constant i.e., $l'A' = lA$, As $l' = 2l$, hence $A' = A/2$

$$\therefore \quad R' = \rho\frac{l'}{A'} = \rho\frac{2l}{A/2} = 4\left(\frac{\rho l}{A}\right) = 4R$$

$= 4 \times 1.7 = 6.8$ ohms.

Example 6:

The temperature coefficient of resistance of a wire is 0.00125°c⁻¹. At 300 K its resistance is one ohm. At what temperature the resistance of wire will be 2 Ω.

Solution:

Temperature 300 K = 27°C. As per question

$$R_{300} = R_0 (1 + \alpha \times 27) = 1$$

and $R_t = R_0 (1 + \alpha \times t) = 2$

$$\therefore \quad \frac{1 + 27\alpha}{1 + t\alpha} = \frac{1}{2}$$

or $2 + 54\,\alpha = 1 + t\,\alpha$

or $2 + 54 \times (0.00125) = 1 + (0.00125)\,t$

On solving, $t = 854°C = 854 + 273$

$= 1127$ K.

Example 7:

A 10 m long wire of resistance 20 Ω is connected in series with a battery of e.m.f. 3V (negligible internal resistance) and a resistance of 10 Ω. Find the potential gradient along the wire.

Solution:

Current, $$I = \frac{3}{10 + 20}$$

$$= \frac{1}{10} = 0.1 \text{ A}$$

Pot. diff. across 20 Ω = 0.1 × 20 = 2.0 V

Potential gradient = 2/10 = 0.2 V m^{-1}.

Example 8:

Twelve cells each having the same e.m.f. are connected in series and are kept in a closed box. Some of the cells are connected in reverse order. The battery is connected in series with an ammeter, an external resistance R and two cells of the same type as in the battery. The current when they aid each other is 3 ampere and is 2 ampere when the two oppose each other. How many cells are connected in reverse order?

(Roorkee 1993)

Solution:

If one cell is connected in reverse order, then it will destroy one more cell. If n cells are connected in reverse order, then 2 n cells will be destroyed. If E is the e.m.f. of each cell, then effective e.m.f. of 12 cells will be = (12 – 2n) E.

This battery if supports two similar cells in series, then supplies a current of 3 A to resistance R. Therefore,

$$3 = \frac{(12 - 2n)\,E + 2E}{R} = \frac{(14 - 2n)\,E}{R} \qquad \text{...(i)}$$

If battery and two cells oppose each other, then the current in the external resistance is 2A. Hence

$$2 = \frac{(12 - 2n)\,E - 2E}{R} = \frac{(10 - 2n)\,E}{R} \qquad \text{...(ii)}$$

Dividing (i) by (ii), we get

$$\frac{3}{2} = \frac{14 - 2n}{10 - 2n} \quad \text{or} \quad n = 1$$

Example 9(a):

1 m long metallic wire is broken into two unequal parts P and Q. P part of the wire is uniformly extended into another wire R. Length of R is twice the length of P and the resistance of R is equal to that of Q. Find the ratio of the resistance of P and R and also the ratio of length of P and Q.

Solution :

L = 1 m, Let length of part P = x

∴ Length of part Q = L – x

P part of wire is uniformly extended to another wire R, having length double of P i.e. length of extended wire = 2x

As volume will remain constant during extension hence, xA = 2x × A' i.e. $A' = \frac{A}{2}$ = c.s. area of wire R

Now, Resistance of $R = \rho.\frac{2x}{A'} = \rho.\frac{4x}{A}$

and Resistance of $Q = \rho.\frac{(L-x)}{A}$

As, resistance of R = Resistance of Q

$\therefore\ \rho.\frac{4x}{A} = \frac{\rho(L-x)}{A}$

or 4x = L − x

or 5x = L

or $x = \left(\frac{L}{5}\right)$

∴ Length of $P = \left(\frac{1}{5}\right)$ m

and Length of $Q = \left(\frac{4}{5}\right)$ m

∴ Ratio of lengths of P and Q = 1 : 4

$$\frac{\text{Resistance of P.}}{\text{Resistance of R}} = \frac{\rho.\frac{x}{A}}{\rho.\frac{4x}{A}} = \frac{1}{4}$$

or **1 : 4**

Example 9(b):

A n type silicon sample of width 4×10^{-3} m, thickness 25×10^{-5} m and length 6×10^{-2} m carries a current of 4.8 mA when the voltage is applied across the length of the sample. What is the current density ? If the free electron density is 10^{22} m^{-3} then find how much time does it take for the electrons to travel the full length of the sample?

Solution :

Given that $b = 4\times10^{-3}$ m, $d = 25\times10^{-5}$ m

$l = 6 \times 10^{-2}$ m, i = 4.8 m amp, n = $10^{22}/m^3$.

current density,

$$j = \frac{i}{A} = \frac{i}{bd} = \frac{4.8 \times 10^{-3}}{4 \times 10^{-3} \times 25 \times 10^{-5}}$$

$= \mathbf{4.8 \times 10^3}$ **amp/m²**.

Now, $j = n\,ev_d$

$$\therefore \quad v_d = \frac{j}{ne} = \frac{4.8 \times 10^3}{10^{22} \times 1.6 \times 10^{-19}} = 3 \text{ m/s}$$

$$\therefore \quad t = \frac{l}{v_d} = \frac{6 \times 10^{-2}}{3} = 2 \times 10^{-2} \text{ sec.}$$

Example 9(c):

A series battery of 6 cells each of e.m.f. 2V and internal resistance 0.5 Ω is charged by a 100 V d.c. supply. What resistance should be used in the charging circuit in order to limit the charging current to 8A. Using this relation, obtain (a) the power supplied by the d.c. source, (b) the power dissipated as heat and (c) the chemical energy stored in the battery in time 15 minutes.

Solution:

Given, no. of cells n = 6;

E.M.F. of each cell, E = 2V;

internal resistance of each cell, r = 0.5 Ω;

charging voltage V = 100 V.

Let R be the resistance used in the series of the circuit while charging the cells. Then current in the circuit will be

$$i = \frac{V - nE}{nr + R}$$

or
$$R = \frac{V - nE}{i} - nr$$

$$= \frac{100 - 6 \times 2}{8} - 6 \times 0.5$$

$$= 11 - 3 = 8\ \Omega$$

(a) Power supplied by d.c. source

$= V \times i = 100 \times = 800$ W

(b) Power dissipated as heat

$= i^2 (R + nr)$

$= 8^2 (8 + 6 \times 0.5)$

$= 704$ W

(c) Rate at which the chemical energy is stored
$= 800 - 704 = 96$ watt

$\therefore$ Chemical energy stored in 15 min.

$= 96 \times 10 \times 60 = 86400$ J.

Example 10:

Two uniform wires of same material each weighting 1 g but one having double the length of the other, are connected in series, carrying a current of 10A. The length of longer wire is 20 cm. Calculate the rate of consumption of energy in each of the two wires. Which one wire gets hotter? Density of the material of wire = 11 g/cc, specific resistance of the material = 20×10^{-5} Ω cm.

Solution:

Let a_1 and a_2 be the area of cross-section of shorter and longer wire respectively.

As, mass = volume × density = $al\rho$

$\therefore \quad 1 = a_1 \times 10 \times 11 = a_2 \times 20 \times 11$

or $\quad a_1 = \dfrac{1}{10 \times 11}$ cm^2 and $a_2 = \dfrac{1}{20 \times 11}$ cm^2

$$\therefore \quad R = 20 \times 10^{-5} \times \frac{10}{1/(10 \times 11)}$$

$$= 20 \times 10^{-5} \times 10 \times 10 \times 11$$

$$= 22 \times 10^{-2}\ \Omega$$

$$\therefore \quad R_2 = 20 \times 10^{-5} \times \frac{20}{1/(20 \times 11)}$$

$$= 88 \times 10^{-2}\ \Omega$$

And rate of heat produced

$$H_1 = I^2R_1$$
$$= (10)^2 \times 22 \times 10^{-2} = 22 \text{ W}$$

and $H_2 = I^2R_2 = (10)^2 \times 88 \times 10^{-2} = 88$ W

Thus the wire of longer length gets hotter.

Example 11:

An electric coil is connected in series with a resistance of x Ω across the 240 volt, mains, the coil being immersed in 1 kg of water at 20°C. The temperature of water rises to boiling point (i.e. 100°C) in 10 minutes. When a second heating experiment is made with the resistance x short circuited, the time required to develop the same quantity of heat is reduced to 6 minutes. Calculate the value of x.

Solution:

Case (i) Let R be the resistance of heating coil. Then current through heating coil will be:

$$I = \frac{240}{X + R}.$$

Therefore, $$\left(\frac{240}{X+R}\right) R = \frac{1 \times 4.2 \times 10^3 \times (100 - 20)}{10 \times 60}$$

$$= \frac{4.2 \times 10^3 \times 80}{10 \times 60} \quad \text{...(i)}$$

Case (ii)

$$\frac{(240)^2}{R} = \frac{4.2 \times 10^3 \times (100 - 20)}{6 \times 60} \quad \text{...(ii)}$$

On solving (ii), we get

$$R = 432/7 \ \Omega$$

Putting this value in (i) and solving it, we get

$$x = 17.83 \ \Omega.$$

Example 12:

A metal wire of diameter 2 × 10^{-3} m and length 100 m has a resistance of 0.5475 ohms at 20° C and 0.805 ohms at 150° C. Find the values of

(i) temperature coefficient of resistance

(ii) its resistance at 0° C

(iii) its resistivities at 0° C and 20° C

Solution :

If R_{20} and R_{150} are the resistances of the wire at 20° C and 150° C respectively and α is the temperature coefficient of resistance of the metal, then

$R_{20} = R_0 (1 + \alpha \times 20) = 0.5475$ ...(1)

$R_{150} = R_0 (1 + \alpha \times 150) = 0.8050$ ...(2)

Dividing equation (1) by equation (2)

$$\frac{1+20\alpha}{1+150\alpha} = \frac{0.5475}{0.8050}$$

$$\therefore \ \alpha = \frac{0.8050 - 0.5475}{0.5475 \times 150 - 0.8050 \times 20} = \mathbf{3.9 \times 10^{-3}\,{}^\circ C}$$

(ii) Putting value of α in equation (1), we get

$$R_0 = \frac{0.5475}{1+\alpha \times 20} = \frac{0.5475}{1+3.9\times 10^{-3} \times 20}$$

= 0.5079 ohms.

(iii) Now $R_0 = \rho_0 (l / A)$

$$\therefore \ \rho_0 = \frac{R_0 A}{l} = \frac{0.5079 \times \pi \times 10^{-6}}{100}$$

$= 1.596 \times 10^{-8}$ ohms × meter

and $\rho_{20} = \rho_0 (1 + \alpha \times 20) = 1.596 \times 10^{-8}[1 + 3.9 \times 10^{-3} \times 20]$

$= 1.720 \times 10^{-8}$ ohms × meter

Example 13:

Calculate the following for a 100 W – 250 V bulb.

(i) Resistance of the filament of the bulb.

(ii) Current through bulb when it is connected to 250 volt line.

(iii) Power consumed when bulb is connected to a 200 volt line.

Solution :

$$\text{(i)} \ \ R = \frac{V^2}{P} = \frac{250 \times 250}{100} = \mathbf{625\ ohms}$$

(ii) Current through bulb

$$i = \frac{P}{V} = \frac{100 \text{ watt}}{250 \text{ Volt}} = \mathbf{0.4\ amp.}$$

(iii) Power consumed when connected to a 200 volt line

$$P' = \frac{V^2}{R} = \frac{200 \times 200}{625} = \mathbf{64\ Watt}$$

Example 14(a):

One Kilowatt electric heater is to be used with 220 volt DC supply.

(i) What is the current in the heater ?

(ii) What is its resistance ?

(iii) What is the power dissipated in the heater ?

(iv) How much heat in calories is produced/sec?

(v) How many grams of water at 100 ° C will be converted per minute into steam at 100 ° C with the heater ? Assume that the heat losses due to radiaion are negligible. Latent heat of steam = 540 cal/gram.

Solution :

P = 1000 Watt, V = 220 volts.

(i) Current in the heater i = (P/V) = (1000 W/220 V) = 4.55 A

(ii) Resistance of the heater element

$$R = \frac{V^2}{P} = \frac{220 \times 220}{1000} = \mathbf{48.4\ ohms}$$

(iii) Power consumed by the heater = 1000 Watt

(iv) Heat produced per sec.

$$H = \frac{P \times t}{J} = \frac{1000 \times 1}{4.2} = \mathbf{240\ cal/sec}$$

(v) Heat produced in one minute

= 240 × 60 ; Q = 14400 cal

Hence Q = ML

i.e. Mass of water at 100° C converted into steam at 100° C

= M = Q/L= (14400 / 540)

= 26.8 gms.

Example 14(b):

Two Bulbs rated at 60 W – 240 V and 100 W – 200 V, are connected in series across a 240 volt line. If the resistance of two bulbs remain constant, what will be the potential difference across each bulb ?

Solution :

Resistance of 1st bulb $R_1 = \frac{V^2}{P_1} = \frac{240 \times 240}{60} = 960$ ohms

Resistance of 2nd bulb $R_2 = \frac{V^2}{P_2} = \frac{200 \times 200}{100} = 400$ ohms

As two bulbs are connected in series, hence their total resistance

$R = R_1 + R_2 = 960 + 400 = 1360$ ohms.

Note : We will study in next chapter, that effective resistance, when two or more resistance are connected in series, is given by : $R = R_1 + R_2 + ...$

Hence current flowing through each resistance (current remains same in series combination of resistances)

$$i = \frac{V}{R_1 + R_2} = \frac{240}{1360} \text{ amp}$$

$\therefore$ V_1 = potential difference across 1st bulb

$$= iR_1 = \frac{240}{1360} \times 960 = 169.4 \text{ Volts}$$

and V_2 = potential difference across 2nd bulb

$$= iR_2 = \frac{240}{1360} \times 400 = 70.6 \text{ Volts}$$

Example 14(c):

Two tungston lamps with resistances R_1, R_2 respectively at full incandescence are connected, first in parallel and then in series, in a lighting circuit of negligible internal resistance. If $R_1 > R_2$,

(i) which lamp will glow more brightly when they are connected in parallel ?

(ii) If the lamp of resistance R_1 now burns out, how will the illumination produced change ?

(iii) *Which lamp will glow more brightly when they are connected in series ?*

(iv) *If the lamp of resistance R_2 now burns out, and the lamp of resistance R_1 alone is plugged in, will the illumination increase or decrease?*

(v) *In any case, would physically bending a supply wire cause any change in the illumination?*

Solution :

(i) when the lamps are connected in parallel, then potential difference V across each lamp will be same and will be equal to potential difference necessary for full brightness of each bulb.

Because illumination produced by a lamp is proportional to electric power consumed in it, and power consumed

$$P_1 = \frac{V^2}{R_1} < \frac{V^2}{R_2} = P_2$$

hence *illumination produced by 2nd bulb will be higher than produced by 1st bulb i.e. bulb having lower resistance will shine more brightly.*

(ii) When R_1 burns out, then power is dissipated in R_2 only. Because internal resistance is quite low in lighting circuit, potential difference is still equal to V, hence power dissipated in 2nd lamp i.e.

$$\frac{V^2}{R_2} < \left(\frac{V^2}{R_1} + \frac{V^2}{R_2}\right) \text{ i.e. net power consumed initially.}$$

In other words, *net illumination will now decrease.*

(iii) When the two lamps are connected in series, the potential difference across each lamp will be different but current i flowing through each lamp will be same.

Hence $P_1 = i^2 R_1 > i^2 R_2 = P_2$

i.e. illumination produced by 1st lamp will be higher as compared to that produced by 2nd lamp i.e. *lamp having higher resistance will glow more brightly.*

(iv) When lamp of resistance R_2 burns out and only lamp of resistance R_1 is connected in the circuit, then current flowing through the circuit will change. Let new current is i.

Because potential difference still remains same (due to low internal resistance), hence

$i' R_1 = i(R_1 + R_2)$

or $i' = [i (R_1 + R_2)]/R_1$

If P' is the power consumed, then

$P' = i'^2 R_1 = [i^2 (R_1 + R_2) . (R_1 + R_2)]/R_1$

When both the lamps were present then total power consumed was given by

$P_S = P_1 + P_2 = i^2 (R_1 + R_2)$ i.e. $P' > P_S$

i.e. *Illumination gets increased when only one bulb is used.*

(v) If a water pipe is given a bend at some point, then it definitely reduces the flow of water in the pipe but this is not true in case of an electric current flowing in a conductor because electric current is established in a conductor due to drift motion of electron in it along the line of the potential gradient. Hence illumination is not effected due to bending along the length of supply wires.

Example 15:

An electric kettle rated at (1500 W - 240 V) takes 7 minutes to increase the temperature of 1.8 kg of water from 20° C to100 ° C.

(a) determine the efficiency of the kettle and

(b) find out the mass of water which may be boiled by 1KWH of electrical energy.

The specific heat capacity of water is 4.18×10^3 Joules/kg/° K

Solution :

Energy supplied to kettle i.e. input Energy

$= P \times t = 1500 \times 7 \times 60 = 630 \times 10^3$ J

Energy obtained from kettle i.e. output energy = $ms.d\theta$

$= 1.8 \times 4.18 \times 10^3 \times (100 - 20) = 602 \times 10^3 J$

(a) Hence efficiency of kettle

$$\eta = \frac{\text{Output Energy}}{\text{Input Energy}} \times 100\% = \frac{602}{630} \times 100 = \mathbf{95.5\%}$$

(b) As Energy required to boil 1.8 kg of water is 630×10^3 J. Hence mass of water boiled by 1 KWH of energy

$$= \frac{1.8 \times 3.6 \times 10^6}{630 \times 10^3} \text{ Kg} = \mathbf{10.29\ Kg}.$$

Example 16:

A generator of e.m.f. 80 volt has an internal resistance of 0.04 ohms. If the terminal voltage is 75 volts, calculate

(i) the current

(ii) power supplied

(iii) power dissipated in the generator

(iv) reading of an ammeter of resistance 1 ohm connected in the circuit.

Solution :

(i) $i = \frac{E - V}{r} = \frac{80 - 75}{.04} = 125$ amp.

(ii) Power suppled P = Vi = 125 × 75 = **9375 Watt**

(iii) Power dissipated P' = $i^2 r = (125)^2 \times 0.04 = 625$ Watt

(iv) Resistance $R = \frac{V}{i} = \frac{75}{125} = 0.6$ ohms

Total external resistance of the circuit R'= 0.6 | 1 = 1.6 ohms

∴ current in the circuit or Reading of ammeter

$$i' = \frac{E}{R' + r} = \frac{80}{1.6 + 0.04} = \frac{80}{1.64} = 48.78 \text{ Amp}$$

Example 17:

A fuse of lead wire has an area of cross section 0.2 mm². On short circuiting, the current in the fuse wire reaches 30 amp. How long after the short circuiting, will fuse begin to melt ? For lead, specific heat 0.032 cal / gm/° C, melting point = 327 ° C, density = 11.34 gm/cm³ and resistivity = 22 × 10⁻⁶ ohms-cm. The initial temperature of wire is 20° C. Neglect heat losses.

Solution :

Given that, Resistivity $\rho = 22 \times 10^{-6}$ ohms-cm = 22×10^{-8} ohms-m;

$A = 0.2\ mm^2 = 0.2 \times 10^{-6}\ m^2 = 2 \times 10^{-7}\ m^2$;

$(\theta_2 - \theta_1) = 327 - 20 = 307° C$;

$d = 11.34$ gms/cm^3 $= 11.34 \times 10^3$ kg/m^3 current $i = 30$ amp, specific heat $s = 0.032$ cal/gm – ° C = 0.032 K cal/kg- ° C.

For the melting of fuse wire,

$$\begin{bmatrix}\text{Heat produced due to} \\ \text{flow of current in fuse} \\ \text{wire in time t.}\end{bmatrix} = \begin{bmatrix}\text{Heat required for increasing} \\ \text{the temperature of fuse wire} \\ \text{upto melting point.}\end{bmatrix}$$

$$\frac{i^2Rt}{4.2} = Q = ms\,(\theta_2 - \theta_1)$$

or $$\frac{i^2\,(\rho l/A)\,t}{4.2 \times 10^3} = (A/d)\; s\; (\theta_2 - \theta_1)$$

$$\left[\frac{(\text{amp})^2 \times [\text{ohm} \times \text{meter} \times (\text{meter/meter}^2)]\,\text{sec}}{4.2 \times 10^3}\right.$$

$$\left. = \text{meter}^2 \times \text{meter} \times \frac{\text{kg}}{\text{meter}^3} \times \frac{\text{K cal}}{\text{kg} - {}^\circ\text{C}} \times {}^\circ\text{C}\right]$$

[K cal = K cal]

Thus, $$\frac{i^2\rho t}{4.2 \times 10^3\,A} = A\;d\;s\;(\theta_2 - \theta_1)$$

$$\therefore\; t = \frac{4.2 \times 10^3\,A^2\,ds(\theta_2 - \theta_1)}{i^2\rho}$$

$$= \frac{4.2 \times 10^3 \times (2 \times 10^{-7})^2 \times (11.34 \times 10^3) \times (0.032) \times 307}{(30)^2 \times 22 \times 10^{-8}}$$

= 0.095 seconds

Example 18:

Forty electric bulbs are connected in series across a 220 V supply. After one bulb is fused, the remaining 39 are connected again in series across the same supply. In which case will there be more illumination and why ?

Solution :

If R is the resistance of each bulb, then total resistance of 40 bulbs in series will be 40R ohms. When connected across a supply voltage V, the total power consumed by the system or illumination produced is given by $P_{40} = V^2 / 40R$

When one of the bulbs is fused, the resistance of the remaining 39 bulbs in series = 39R and power consumed by the system in now given by $P_{39} = V^2/39R$

Thus it is clear that $\frac{V^2}{30R} > \frac{V^2}{40R}$

i.e. illumination produced by 39 bulbs will be higher.

Example 19:

Two uniform wires of the same material, each weighing 1 gm but one having double the length of the other, are connected in series. They carry a current of 10 A. The length of longer wire is 20cm. Calculate the rate of consumption of energy in each of the two wires. Which one will get hotter ? Density of the material of the wire = 11.34 gm/cm³ . Specific resistance of the material = 20.6 × 10⁻⁵ ohms × cm.

Solution :

Because mass and density of each wire are same, hence volume of each wire

$V = (1.0/11.34)\ cm^3$

If A represents the area of cross section of the longer wire of length 20cm, then

$$20\ A = \left[\frac{1.0}{11.34}\right]$$

or $$A = \frac{1}{20 \times 11.34}\ cm^2$$

Hence area of cross-section of the shorter wire = 2A

(C.S. area is doubled because length of shorter wire is half the length of longer wire). Now Resistance of longer wire.

$$R_1 = \rho.\frac{l}{A} = \frac{(20.6 \times 10^{-6}) \times 20}{I} \times (20 \times 11.34)$$

= 0.09344 ohms

∴ Power consumed by the longer wire

$P_1 = i^2R_1 = (10)^2 \times (.09344) =$ **9.344 Watt.**

Resistance of the shorter wire

$$R_2 = \rho \frac{(l/2)}{2A} = \frac{1}{4}\frac{(\rho l)}{A} = \frac{R_1}{4}$$

∴ Power consumed by the shorter wire

$$P_2 = i^2 R_2 = i^2 \frac{R_1}{4} = \frac{P_1}{4} = \frac{9.344}{4} = \mathbf{2.34\ Watt}$$

i.e. $P_2 < P_1$ i.e. longer wire becomes hotter.

Example 20:

A standard 50 W electric lamp in series with a room heater is connected across the mains. If the 50 W bulb is replaced by a 100 W bulb, will the heater output be larger, smaller or remain the same?

Solution :

We know that Resistance of 50 W bulb > Resistance of 100 W bulb ($\because R \propto 1/P$), hence when 50 W bulb is replaced by 100 W bulb in the circuit, the total resistance of the circuit is decreased and a higher current now flows in the circuit i.e. power consumed ($= i^2 R$) by heater is increased or *heater output is increased.*

Example 21:

What current must be passed through a 40Ω immersion heater, dipped in a tank of heat capacity 630 J/K, containing 2.75 kg of water, so that heat developed increases the temperature of the tank from 10°C to 40°C in five minutes. Assume that 4200 Joules of work is equivalant to 1k cal.

Solution :

Because heat capacity of tank or ms = 630 Joules/ °K = 630 Joules/ °C = (630/4200) kcal/°C, = 0.15 k cal/°C, hence quantity of heat required to raise its temperature by 1°C = 0.15 k cal which is just equal to the quantity of heat required to raise the temperature of 0.15 kg of water by the same amount i.e.1°C.

Hence, effective mass of water = 2.75 + 0.15 = 2.9 kg.

and heat required to raise its temperature by 30°C = Q = ms.dθ

= 2.9 kg × (1 k cal/kg – °C) × 30°C

= 2.9 × 30 × k cal = 2.9 × 30 × 4200 Joules

If i is the current which passes through immersion heater, then $i^2 Rt = Q$

$$\therefore \quad i^2 = \frac{Q}{tR} = \frac{2.9 \times 30 \times 4200}{5 \times 60 \times 40} = 30.45$$

Hence i = **55.18 amp**

Example 22

A two meter long wire of resistance 4 ohms and diameter 0.64 mm is coated with plastic insulation of thickness 0.06 mm. When a current of 5 amp is passed through the wire, what will be temperature difference across the insulation in steady state. Coefficient of thermal conductivity of the insulating material = 0.16 × 10^{-2} cal/sec/cm/^{o}C and 1 cal = 4.2J

Solution :

Given that Radius of the metal wire = R = 0.32 mm = .032 cm, Thickness of insulating material = d = .06 mm = .006 cm, Inner radius of insulation = R = .032 cm, Outer radius of insulation = R + d = 0.032 + .006 = .038 cm, Average radius of insulation = r = $\left(\frac{1}{2}\right)$ (.038 + .032) = .035 cm

Curved surface area, corresponding to average radius = A = 2πrL (where L is the length of wire) Heat energy that passes radially through insulating material, may be assumed as passing through curved surface area only because base area is negligible as compared to curved surface area.

In steady state :

$$\left.\begin{array}{l}\text{Heat produced per sec in the metallic}\\ \text{wire due to flow of current}\end{array}\right] = \left[\begin{array}{l}\text{Heat conducted per second radially}\\ \text{through insulating material}\end{array}\right.$$

$$\frac{i^2R}{4.2}\text{ cal} = \frac{KA(\theta_2 - \theta_1)}{a}$$

$$\therefore \quad (\theta_2 - \theta_1) = \frac{i^2Rd}{4.2\,KA} = \frac{i^2Rd}{4.2K \times 2\pi rL}$$

$$= \frac{(5)^2 \times 4 \times .006}{4.2 \times (0.16 \times 10^{-2}) \times (2\pi \times .035 \times 200)}$$

$$= \frac{23.8 \times .006}{0.16 \times 10^{-2} \times 44} = \mathbf{2^{o}C}$$

Example 23:

Two bulbs rated 25 W - 220 V and 100 W - 220 V are connected in series to a 440 volt supply. Show with necessary calculations, which bulb, if any, will fuse ? What will happen if the two bulbs are connected in parallel to the same supply ?

Solution :

Given that P_1 = 25 Watt, P_2 = 100 Watt and voltage V = 220 volt.

Hence, resistances of the filaments of two bulbs are given by

$R_1 = (V^2/P_1) = (220 \times 220/25) = 1936$ ohms

& $R_2 = (V^2/P_2) = (220 \times 220/100) = 484$ ohms

Now, maximum currents that can be allowed to pass through the two bulbs, are given by

i_1 (max) = (P_1/V) = (25/220) = 0.114 amp

i_2 (max) = (P_2/V) = (100/220) = 0.454 amp

Case I : *When bulbs are connected in series,* then total resistance $R_5 = R_1 + R_2$ = 2420 ohms, and current in the circuit = current through each bulb = $i_S = (V'/R_S)$ = (440/2420) = 0.18 **amp.**

As, $i_S > i_1$ (max) but < i_2 (max), hence *bulb rated 25 W – 220 V, will fuse.*

Case II : When bulbs are connected in parallel, then potential difference across each bulb will be same and equal to 440 volt.

Hence current through 25 Watt bulb

$i_1 = \frac{V'}{R_1} = \frac{440}{1936}$ = 0.23 amp & that through 100 Watt bulb

$i_2 = \frac{V'}{R_2} = \frac{440}{484}$ = 0.91 amp

As $i_1 > i_1$ (max) and i_2 (max), hence *both bulbs will fuse.*

Example 24(a):

It is required to send a current of 10 amp through a resistance R = 3Ω.

(a) *What is the minimum number of cells required if each cell has an e.m.f. of 10 V and internal resistance 1 ohm.*

(b) *Find the power dissipated in R.*

Solution :

(a) Suppose N be the required number. Let us arrange in m rows each containing n cells in series. Then mn = N.

The internal resistance of the battery formed

$= \frac{nr}{m} = \frac{n}{m}$ and e.m.f. of the battery = nE = 10n

Then, $10 = \frac{10n}{\frac{n}{m}+3}$

or, $10 = \frac{10\,mn}{n+3m}$ or, n + 3m = mn

or, $n+3\frac{N}{n}$ or, $n^2 + 3N = Nn$

Differentiating with respect to n,

$$2n+3\frac{dN}{dn} = \frac{dN}{dn} \times n + N$$

When N is minimum $\frac{dN}{dn} = 0$,

Hence $n = \frac{N}{2}$

Hence, when N is minimum, $\frac{N^2}{4} + 3N = \frac{N^2}{2}$

or, $\frac{N^2}{4} = 3N$ or, $\mathbf{N = 12}$

Example 24(b):

Two coils when joined in series have an effective resistance of 18 ohms but when joined in parallel, an effective resistance of 4 ohms. Calculate the values of their resistances.

Solution :

If R_1 and R_2 are the resistances of two coils, then in series combination their effective resistance

$R_S = R_1 + R_2$ i.e. $R_1 + R_2 = 18$ ...(1)

In parallel combination, their effective resistance

$$R_P = \frac{R_1R_2}{R_1+R_2} \text{ i.e. } \frac{R_1R_2}{R_1+R_2} = 4$$

or $R_1R_2 = 72$ ...(2)

From eq. (1) and (2)

$R_1 - R_2 = \sqrt{[(R_1 + R_2)^2 - 4R_1R_2]}$

$= \sqrt{[(18)^2 - 4 \times 72]} = 6$ ...(3)

From eq. (1) and (3) :-

R_1 = 12 ohms & R_2 = 6 ohms.

Example 25:

In a silver platinum system, an electrolysis current of 5.0 A is used for a certain time and 0.5 mole of silver is deposited. How many moles of copper and iron will be deposited in their respective plating system if an electrolysis current of 10.0 A passed for twice the time for silver plating. Relative atomic mass of silver = 107.3, of copper = 63.54, of iron = 55.85.

Solution:

$$m = zIt = \left(\frac{M}{Npe}\right) It$$

or $$\frac{m}{M} = \frac{It}{Npe}$$ ...(i)

where M = mass of a mole,

N = Avogadro's number,

p = valence and

e = charge on electron.

The value of p for silver, copper and iron is 1, 2, and 3 respectively. In case of silver m/M = 0.5

From (i), $$0.5 = \frac{5 \times t}{N \times 1 \times e}$$

or $$\frac{t}{Ne} = \frac{0.5}{5} = 0.1$$

In case of copper,

$$\frac{m}{M} = \frac{10 \times 2t}{N2e} = 10 \times 0.1$$

$$= 1 \text{ mole.}$$

In the case of Iron,

$$\frac{m}{M} = \frac{10 \times 2t}{N\,3e}$$

$$= \frac{20}{3} \times 0.1 = \frac{2}{3} \text{ mole.}$$

Example 26:

A series battery of 6 lead accumulators each of e.m.f. 2.2 V and internal resistance 0.50 Ω is charged by a 100 V d.c. supply. What series resistance should be used in the charging circuit in order to limit the current to 7.84? Using the required resistor, obtain

(a) *the power supplied by the d.c. source.*

(b) *the power dissipated as heat.*

(c) *the chemical energy stored in the battery in 15 min.*

Solution:

Total resistance

$= R + nr$

$= R + 6 \times 0.5$

$= R + 3$

Total e.m.f. of the battery

$= 6 \times 2.2 = 13.2$ V

Effective potential difference in the circuit

$= 100 - 13.2 = 86.8$ VV

$$\text{Now, current} = \frac{\text{effective pot. diff.}}{\text{total resistance}}$$

$$\therefore \qquad 7.8 = \frac{86.8}{R + 3}$$

$$\text{or} \qquad R = 8.13\ \Omega$$

(a) Power supplied by d.c. source

$$= VI = 100 \times 7.8 = 780 \text{ watt.}$$

(b) Power dissipated as heat

$$= I^2 (R + nr) = (7.8)^2 (8.13 + 3)$$

$$= 677.15 \text{ watt.}$$

(c) Chemical energy stored in the battery in 15 minutes.

$$= (780 - 677.15) \times 15 \times 60 = 92565 \text{ J}$$

Example 27:

A tangent galvanometer is connected in series with a copper voltameter and a steady current is passed through them. The galvanometer reading is 45° and 0.2988 gram of copper is deposited in half an hour. Find the reduction factor of tangent galvanometer. Given that E.C.E. of copper is 0.00032 gC^{-1}.

Solution:

Let I be the current in the circuit

$$\therefore \quad I = K \tan \theta = \frac{m}{zt}$$

$$\text{or} \quad K = \frac{m}{zt \tan \theta}$$

$$= \frac{0.2988}{0.000332 \times (30 \times 60) \times \tan 45°} = 0.5\text{A}$$

Example 28:

A steady potential difference of 1.62 V is maintained across two platinum electrodes placed in a solution of $CuCl_2$. At the end of 600 s, the mass of copper deposited on the cathode is measured to be 5.92 g. The back e.m.f. of the voltameter is given to be 1.34 V. Estimate the resistance of voltameter. Faraday constant = 96500 C mol^{-1}, relative atomic mass of copper = 63.5.

Solution:

Here, $V = 1.62$ V;

$t = 600$ s;

$m = 5.92$ g

Back e.m.f., $V_b = 1.34$ V;

$$R = ?;$$

$$F = 96500 \text{ C mol}^{-1};$$

Chemical equivalent of copper,

$$E = \frac{63.5}{2} = 31.75$$

As $$m = zIt = \frac{E}{F}It.$$

or $$I = \frac{mF}{Et} = \frac{5.92 \times 96500}{31.75 \times 600}$$

$$= 29.99\text{A}$$

Now, $$R = \frac{V - V_b}{I} = \frac{1.62 - 1.34}{29.99}$$

$$= 9.34 \times 10^{-3}\ \Omega$$

Example 29:

In a house having 220 V-line, the following appliances are operating: (i) 60 W bulb (ii) a 1000 W heater and (iii) a 40 W radio. Calculate (a) the current drawn by heater and (b) the current passing through fuse for this line.

Solution:

Here, $V = 200$ V;

$P_1 = 60$ W;

$P_2 = 1000$ W;

$P_3 = 40$ W.

(a) Current drawn by heater

$$= \frac{P_2}{V} = \frac{1000}{220} = \frac{50}{11}\text{A}$$

(b) Current drawn by bulb

$$= \frac{P_1}{V} = \frac{60}{220} = \frac{3}{11}\text{A}$$

Current drawn by Radio

$$= \frac{P_3}{V} = \frac{40}{220} = \frac{2}{11}\text{A}$$

Current passing through fuse for the line

$$= \frac{50}{11} + \frac{3}{11} + \frac{2}{11} = 5A.$$

Example 30:

A piece of metal weighing 200 gms is to be electroplated with 5% of its weight in gold. If the strength of the available current is 2 amperes, how long would it take to deposit the required amount of gold, E.C.E. of H = 0.1044 × 10^{-4} g./C, atomic weight of gold = 197.1 and atomic weight of hydrogen = 1.008.

Solution :

Mass of gold to be deposited, m = [200 × (5/100)] = 10 gms

i = 2 amp, W_2 = chemical equivalent of gold = (197.1/3) = 65.7

W_1 = chemical equivalent of Hydrogen = 1.008

Hence,according to Faraday's 2nd law :

$W_1/Z_1 = W_2/Z_2$.

$\therefore$ W.C.E. of gold = $Z_2 = (W_2/W_1) \times Z_1 = \frac{65.7}{1.008}$ $\times 0.1044 \times 10^{-4}$ g/C

From Faraday's first law : m = Zit

$$\text{or } t = \frac{m}{Zi} = \frac{10}{(65.7/1.008) \times 0.1044 \times 10^{-4} \times 2}$$

$$= \frac{10 \times 1.008}{65.7 \times 0.1044 \times 10^{-4} \times 2}$$

= 7347.9 seconds = **2 hours 2 min 27.9 sec.**

Example 31:

A 3 ton motor operated vehical goes up an inclined plane having a slope of sin^{-1} (1/25), at an uniform speed of 36 km/hour. The tractive resistance may be taken as 40 kgf per ton. Assuming effciency of a motor = 85% and mechanical efficiency between motor and road wheels = 80%, calculate

(i) output of motor, and

(ii) current taken by the motor which works on 240 volts.

Solution :

(i) m = mass of vehicle = $3 \times 1000 = 3000$ kg

Speed of motor = v = 36 km/hour = $36 \times (5/18) = 10$ m/sec.

Tractive resistance = F = $3 \times 40 = 120$ kgf = $120 \times 9.8 = 1176$ N.

Component of the weight of the car along the plane

$= W\sin\theta = 3000 \times 9.8 \times (1/25) = 1176$ N.

As the vehicle is moving upwards, hence tractive resistance will also act down the plane.

Hence total force down the plane = 1176 + 1176 = 2352 N.

$\therefore$ Total power output at road wheels

= Total Force × speed = $2352 \times 10 = 23520$ Watt.

Because this is 80% of motors' output, we have output of motor

= 23520/0.8 = 29400 Watt

(ii) Because efficiency of motor is 85%, hence power input from mains

$$= \frac{29400}{0.85} = 34588 \text{ Watt.}$$

i.e. Vi = 34588,

$$\therefore \quad i = \frac{34588}{V} = \frac{34588}{240} = \mathbf{144\ Amp.}$$

Example 32(a):

Two heater coils made of the same material are connected in parallel across the mains. The length and diameter of the wire of one of the coils are double that of the other. Which one of them will produce more heat?

Solution :

We know that

$$P = \frac{V^2}{R} \text{ and } R = \frac{\rho L}{A},$$

hence, $$P = \frac{V^2 A}{\rho L}$$

Since potential difference applied is same across both wires, hence

$(PL/A) = (V^2/\rho) = \text{constant}$

$$\text{or}, \frac{P_1L_1}{A_1} = \frac{P_2L_2}{A_2}$$

$$\text{or}, \frac{P_1L_1}{\pi r_1^2} = \frac{P_2L_2}{\pi r_2^2}$$

$$\text{or}, \frac{P_2}{P_1} = \frac{L_1}{L_2} \times \frac{r_2^2}{r_1^2} = \frac{L_1}{2L_2}\left(\frac{2r_1}{r_1}\right)^2 = 2$$

or $P_2 = 2P_1$.

i.e. power devoloped in 2nd wire = 2 × power developed in 1st wire

i.e. *2nd wire will develop more heat as compared to that developed by 1st wire in the same time.*

Example 32(b):

A series-parallel combination battery consists of 300 identical cells, each with an internal resistance 0.3 Ω. It is connected to external resistance 10 Ω. Find the number of parallel groups consisting of equal number of cells connected in series, at which the external resistance generated the higher thermal power.

Solution:

Let there be m rows of cells and n cells in each row of cells. Then total number of cells,

$$N = mn \quad \text{or} \quad n = N/m \qquad \text{...(i)}$$

Let E, r be the e.m.f. and internal resistance of each cell and R be the external resistance. Then total internal resistance of all the m rows of cells = nr/m. Total resistance of the whole circuit = (R + nr/m). Total e.m.f. of all the cells = nE.

Current in the external resistance R will be

$$i = \frac{nE}{R + nr/m} = \frac{(N/m)E}{R + (N/m)r/m}$$

$$= \frac{mNE}{m^2R + Nr}$$

Heat generated in resistance R,

$$H = i^2R = \left(\frac{mNE}{m^2R + Nr}\right)^2 R \qquad \text{...(ii)}$$

For H to be maximum,

$$\frac{dH}{dm} = 0.$$

Differentiating (ii) w.r.t. m and equating to zero, we get

$$m = \sqrt{Nr/R} = \sqrt{300 \times 0.3/10} = 3.$$

Example 33:

Power from a 64 V d.c. supply goes to charge a battery of 8 lead accumulators each of e.m.f. 2.0 V and internal resistance 1/8 Ω. The charging current also runs an electric motor placed in series with the battery. If the resistance of the windings of the motor in 7.0 Ω and the steady supply current is 3.5 A, obtain (a) the mechanical energy yielded by the motor, and (b) the chemical energy stored in the battery during charging in 1h.

Solution:

Here, Applied voltage,

$$V = 64 \text{ V}$$

Total e.m.f. of 8 lead accumulators in series,

$$E = 20 \times 8 = 16 \text{ V}$$

Total internal resistance of 8 lead accumulators,

$$r = (1/8) \times 8 = 1\ \Omega$$

Steady current,

$$I = 3.5 \text{ A}$$

Resistance of the windings of the motor,

$$R = 7.0\ \Omega$$

Rate of consumption of energy in the circuit

$$= VI = 64 \times 3.5 = 224 \text{ W}$$

Rate of storing chemical energy in the battery

$$= EI = 16 \times 3.5 = 56 \text{ W}$$

Rate of energy dissipation as heat in the circuit

$= I^2 (R + r) = (3.5)^2 (7 + 1) = 98$ W

Rate of mechanical energy yielded by motor

$= 224 - (56 + 98) = 70$ W

(a) Mechanical energy yielded by motor in $1h$

$= 70 \times 3600 = 252000$ J

(b) Chemical energy stored in the battery in $1h$

$= 56 \times 3600 = 201600$ J.

Example 34:

A copper wire having cross sectional area 0.5 mm² and a length of 0.1 m is initially at 25°C and is thermally insulated from the surroundings. If a current of 10 ampere is set up in this wire

(i) find the time in which the wire will start melting. The change of resistance with temperature of the wire may be neglected.

(ii) What will be the time taken if length of the wire is doubled?

Given for copper wire density $= 9 \times 10^3$ *kg* m^{-3}; *specific heat* $= 9 \times 10^{-2}$ *k cal* kg^{-1} $(°C^{-1})$, *melting point 1075°C, specific resistance* $= 1.6 \times 10^{-8}$ *ohm metre.*

Solution:

(i) Here, $\rho = 1.6 \times 10^{-8}\ \Omega$ m;

$l = 0.1$ m;

$A = 0.5 \times 10^{-6}$ m²

$\therefore$ Resistance of copper wire, $R = \dfrac{\rho l}{A}$

$$= \frac{1.6 \times 10^{-8} \times 0.1}{0.5 \times 10^{-6}} = 0.0032\,\Omega$$

Mass of copper wire,

m = volume × density

$= (Al) \times$ density

$= (0.5 \times 10^{-6} \times 0.1) \times (9 \times 10^3)$

$= 45 \times 10^{-5}$ kg

sp. heat of copper, $c = 0.09$ k cal kg^{-1} $(°C^{-1})$, change in temperature.

$d\theta = 1075 - 25 = 1050°C$

If H is the heat required to melt the copper wire, then

$$\frac{i^2 Rt}{4.2 \times 10^3} = m \times c \times d\theta$$

or
$$\frac{(10)^2 \times 0.0032 \times t}{4.2 \times 10^3}$$

$$= 45 \times 10^{-5} \times 0.09 \times 1050$$

One solving we get, t = 558 seconds.

(ii) When the length of the wire is doubled, its resistance is doubled. Now the mass of the wire is also doubled. Then

$$\frac{i^2 (2R)t}{4.2 \times 10^3} = (2\ m) \times c \times d\theta$$

or
$$t = \frac{mc \times d\theta \times 4.2 \times 10^3}{i^2 R}$$

It means t remains the same = 558 seconds.

Example 35:

An electrically heating coil is placed in a calorimeter containing 360 gram of water at 10°C. Then coil consumes energy at the rate of 90 watts. The water equivalent of the calorimeter and the coil is 40 gram. What will be the temperature of water after 10 minutes?

$J = 4.2\ J\ cal^{-1}$.

Solution:

Heat produced by heating coil

$$= \frac{90 \times 10 \times 60}{4.2} \text{ cal.}$$

Heat taken by water and calorimeter

$$= 360 \times 1 \times (\theta - 10) + 40\ (\theta - 10) \text{ cal}$$

$$\therefore \quad \frac{90 \times 10 \times 60}{4.2} = (360 + 40)\ (\theta - 10).$$

On solving we get, θ = 42.14°C.

Example 36:

A thermocouple is made of iron and nickel and the difference of temperature between their junction is 100° C. Find the thermo e.m.f.

developed across such a couple when the thermo e.m.f. for iron lead couple is 12.5 μ volt/°C and for nickel lead couple is 15μV/°C.

Solution:

Here $E_{Pb}^{Fe} = 12.5\ \mu\ V/°C$;

$E_{Ni}^{Pb} = -\ 15\mu\ V/°C$

According to law of intermediate metals,

$$E_{Ni}^{Fe} = E_{Pb}^{Fe} + E_{Ni}^{Pb} = E_{Pb}^{Fe} - E_{Pb}^{Ni}$$

$$= 12.5 - (-\ 15) = 27.5\ \mu\ V/°C.$$

So thermo e.m.f. for 100°C difference of temperature

$$= 27.5 \times 100 = 2750\ \mu\ V.$$

Example 37:

How many times will be total resistance of n identical conductors be reduced if the series arrangement is changed to parallel one.

Solution:

Effective resistance in series combination:

$$R_s = R + R + ...\ n \text{ terms} = nR.$$

Effective resistance in parallel combination

$$\frac{1}{R_P} = \frac{1}{R} + \frac{1}{R} + ...\ n \text{ terms} = \frac{n}{R}$$

or $R_P = \dfrac{R}{n}$ $\quad \therefore \dfrac{R_p}{R_s} = \dfrac{R/n}{nR} = \dfrac{1}{n^2}$

2

Attracted Disc Machines and Electrometers

INTRODUCTION

The devices used to produce electrical charges on a large scale are called electrostatics machines or generators. In fact, *these machines do not generate the charges but they only separate the charges.* The principle of attraction between the plates of charged capacitor was utilized by Kelvin for constructing the attracted disc-electrometer. In general, there are two types of machines namely.

(1) Frictional machines,

(2) Induction machines.

Now a days the frictional machines are not used.

The commonly used induction machines are:

(1) Electrophorus, divested by Volta in 1775.

(2) Wimshurst machines,

(3) Van de Graaff generator

Here, we shall discuss only the Van de Graaff generator, for other *induction machines see some school books.*

PHYSICAL MEANING OF POLARISATION OF DIELECTRIC

When a dielectric is placed in an external electric field, the electrically charged particles (atoms, ions, molecules) with their displacement arrange is such order that *dielectric acquires a certain electric moment, such a process is called polarisation.*

Linear dielectrics show a *direct proportionality* between the electric moment P (induced moment) acquired by the particle during the process of polarisation and the intensity of the electric field acting on the particle *i.e.,* the local electric field inside the dielectric E_{int}. Mathematically we represent,

$$P \propto E_{int}$$

$$P = \alpha E_{int} \quad \text{...(1)}$$

The coefficient of proportionality α in equation (1) is known as polarizability of the given dipole. Polarisability reflects the properties of an individual particle of matter and not of large volume of it hence it is called microscopic electrical parameter of a dielectric.

If, $E_{int} = 1$, in equation (1)

then, $P = \alpha$ ...(2)

So the *polarisability* α, *is dipole moment per unit field strength.* If there are N dipoles in unit volume of the matter then equation (1) can be written as,

$$P = N \alpha E_{int} \quad \text{...(3)}$$

Here P is the average dipole moment of N dipoles

Unit for Polarisability

When dipole moment P is measured in coul. meter, (P = charge × distance) and E is volt/metre, we get SI unit for α, from equation (1),

$$\alpha = \frac{\text{Coul.meter}}{\text{Volt / meter}} = \frac{\text{Coul.meter}^2}{\text{Volt}}$$

(coul = volt × Farad)

$$\alpha = \text{Farad. meter}^2$$

Capacitors

A capacitor (or condenser) is a device for storing charge. In actual practice, a capacitor is an electrical device consisting of two conductors separated by an insulating or dielectric medium (including air) and carrying equal and opposite charges. The conductors are called pates and may be of any shape. The earliest capacitor was invented almost accidentally by

VAN MUSSCHENBROEK of Leden in about 1746, and became known as a Leden jar.

Parallel Plate Capacitor

As an example let us take two similar flat conducting plates arranged parallel to one another, separated by distance d. Let A be the area of each plate and suppose that there is a charge Q on one plate and –Q on the other. Away from the edge, the field $\vec{E}$ is very nearly uniform in the region between the plates and given by,

$$E = \frac{\sigma}{\epsilon_0} \quad \text{or} \quad \sigma = \epsilon_0 E$$

where σ is the density of the surface charge on the inner surface of the plate. If V is the potential difference between the plates then, $E = V/d$, hence above equation reduces,

$$\sigma = \epsilon_0 \frac{V}{d}$$

If we neglect the variation of $\vec{E}$ near the edge of the plates and therefore of σ, the total charge on one plate represented by Q is given by,

$$Q = \epsilon_0 \frac{VA}{d}$$

or

$$\frac{Q}{V} = \frac{\epsilon_0 A}{d} \qquad ...(4)$$

The ratio $\frac{Q}{V} = \frac{\epsilon_0 A}{d}$, *though it is only approximation in parallel plates, but it is clear that the value of R.H.S. will depend only on the size and geometrical arrangement of the plates.* That is, *for a fixed pair of conductors, the ratio of charge to potential difference will be a constant. We call this constant the capacitance of the capacitor and denoted by C, hence,*

$$\frac{Q}{V} = \text{Constant} = C$$

or

$$Q = CV \qquad ...(5)$$

Thus, the capacity of the parallel plate capacitor, with edge fields neglected, is given by,

$$C = \frac{\epsilon_0 A}{d} \qquad ...(6)$$

Units

From Equation (5) unit of capacitance is coul/volt and is called Farad related by,

$$1\text{Farad} = \frac{1 \text{ coulomb}}{1 \text{ volt}} \qquad ...(7)$$

However, farad is very large unit, we use smaller unit as micro farad or micro-micro farad.

$1 \mu F = 10^{-6} F$ and $1 \mu\mu F = 10^{-12} F$.

CONSTRUCTION OF ELECTROMETERS

It is essentially a guard ring air condenser consisting of two metallic plate A and B and guard ring G to avoid edge effect. The central plate A which can freely pass concentrically through G serves as the attracted disc and is held in position with the help of spring S. Spring S is connected to a rod which can be moved up or down with micrometer screw N. Hence plate A can be moved by screw N.

The plate A and guard ring G are always charged to the same potential and the plate B is either earthed or charged to a different potential. In order to obtain uniform electric field between A and B, the plate A is connected to g by light flexible wire, then the lines of force will be parallel and equidistant.

Theory : Let V_a and V_b be the potentials of plates A and B respectively and d be the distance between them, then the electric field,

$$E = \frac{V_a - V_b}{d} \qquad ...(1)$$

Now the plate A will experience a force of attraction per unit area *i.e.*, pressure given by

$$P = \frac{\varepsilon_o E^2}{2}$$

Therefore, the plate A will experience a force of attraction.

$$F = \frac{\varepsilon_o E^2 A}{2} \qquad ...(2)$$

where A is the "effective area" of the plate which is equal to the actual area of the plate A plus half of the area of the gap between A and G substituting E from Eq. (1) Eq. (2), we get

$$F = \frac{\varepsilon_o}{2}\left(\frac{V_a - V_b}{d}\right)^2 A$$

or $$V_a - V_b = d\sqrt{\frac{2F}{\varepsilon_o A}} \text{ Volts} \qquad ...(3)$$

PROCEDURE FOR MEASURING THE POTENTIAL DIFFERENCE

1. At first the plates A, B and C are connected to the earth so that the potential difference between the plates is zero. A small mass m is then placed on the plate A, due to weight mg, A gets depressed below G with the help of the screw nut N, the plate A is raised to the same place as GG.
2. Now, the small mass m is removed so that the plate A is pulled up by the spring S and is above the place of G. The earth connection of all the plates is now broken.
3. Next the plate A is connected to a potential V_o i.e., the plate is charged to the potential Vo by a machine (Kelvin Replenishes). The plate B is then connected to one of the points, the potential difference between which we wish to measure. The position of B is adjusted with the help of screw M until the plate A comes in level with the guard ring. The reading on the scale is noted, say, R_1 corresponding to the distance between the plates d1. Here, the force of attraction between A and B is mg. If the potential of the first point be V_1, then from Eq. (1)

$$V_o - V_1 = d_1\sqrt{\frac{2mg}{\varepsilon_o A}} \qquad ...(1)$$

4. Finally, the plate B is connected to the second point whose potential is V_2. The position of B is again in adjusted so that A comes in level with the guard ring. Let the reading on scale be R_2 and the distance between the plates be d_2, then

$$V_o - V_2 = d_2 \sqrt{\frac{2mg}{\varepsilon_o A}} \qquad ...(2)$$

Subtracting (2) from (1), we get

$$V_2 - V_1 = (d_1 - d_2) \sqrt{\frac{2mg}{\varepsilon_0 A}} \qquad ...(3)$$

Since $d_2 - d_1 = R_2 - R_1$, we get

$$V_2 - V_1 = R_1 - R_2 \sqrt{\frac{2mg}{\varepsilon_0 A}} \qquad ...(4)$$

Absolute Electrometer

It is called the *"absolute electrometer"* because the potential difference $V_2 - V_1$ in equation (4) is known in terms of the difference between the two micrometer readings, the mass m and the area A. That is the *potential difference $V_2 - V_1$ is obtained in terms of "fundamental quantities" such* as mass and length, which can be measured. No calibration is required in this electrometer, hence it is called an absolute electrometer.

It is, however, *less sensitive* than other electrometers. Further, it requires much setting, therefore, it is more cumbersome to use it than a voltmeter.

Quadrant Electrometer

The attracted disc electrometer, though it gives absolute values is not very *sensitive*. Lord Kelvin devised another more sensitive electrometer known as quadrant electrometer. It was further modified by Dolezalek. The Dolezalek quadrant electrometer possesses *high sensitivity and accuracy.*

Construction : The quadrant electrometer consists of a hollow cylindrical box of brass, 5 cms. in diameter and 1 cm in height, divided into four hollow quadrants, The grandaunts are mounted separately on amber insulating stand. The opposite pairs of quadrants AA and BB are joined together to the terminals T_1 and T_2 respectively can be connected to a source potential Va and Vb. N is a light paddle-shaped van called the meedle which is made of thin paper coated with silver or aluminium. The needle is suspended by a quartz fibre so as to swing freely in a horizontal place without touching the grandaunt. A small mirror M is attached to the suspension fibre to read the deflection by lamp scale arrangement.

Working : When the pairs of quadrants AA and BB are earthed or connected to the same potential, the needle rests symmetrically between them. But when these are charged to different potentials the needle deflects from the higher to the lower potential. This twists the suspension fibre S and a restoring couple acts which bring the needle to rest. *When the deflecting couple due to the applied potential difference becomes equal to the restoring couple* due to the twist, the resulting deflection of the needle, say, θ is proportional to the difference of potential $V_a - V_b$. This is shown in Fig. 3.3 where dotted line indicates the new position of the needle N and θ is deflection. Thus,

$$V_a - V_b \ \theta \propto$$

The portion of needle, which lies between one pair of the quadrants forms a parallel plate condenser, and that which lies between the other pair forms another condenser. When the needle moves through an angle θ from AA quadrants to BB quadrants, the area of needle shown dotted is transferred from AA to BB quadrants. Thus when the needle is deflected, its area in quadrants BB increases and that in quadrants AA decreases.

The increase in area of one face of the needle in BB is area Obb' or Occ'. Let area Obb' or Occ' is S and r be the radius of the needle, then an angle 2π represents an area πr^2 the angle θ will represent an area obtained by the condition, *i.e.*,

$$= \frac{\text{area of Obb}'(S)}{\text{area of the circle } \pi r^2}$$

$$= \frac{\text{angle subtended by the Obb' at centre } \theta}{\text{angle subtended by the circle at centre } 2\pi}$$

$$\text{Hence } S = \frac{\theta}{2\pi} \cdot \pi r^2 = \frac{1}{2} r^2 \theta$$

This is same as the corresponding decrease in the area of one face of needle in AA. As the needle has two arms and two faces, therefore, the total area of the needle transferred from AA to BB

$$= 4. \frac{1}{2} r^2 \theta = 2r^2 \theta$$

If d' is the distance between the needle and the quadrant, the *increase in the capacity of the* condenser formed by the needle and quadrants BB due to this increases in area in given by

$$DC = \frac{\varepsilon_o 2r^2 \theta}{d} \qquad ...(5)$$

since, C = eoA/d,

However, *the capacity of the condenser formed by the needle and quadrants AA will decrease by an "equal amount".* Since the capacity of BN condenser increases while the potential difference ($V_o - V_b$) between N (needle) and B remains constant, it "*gains charge*" whereas the capacity of AN condenser decreases while the potential difference ($V_o - V_a$) between N and A remains constant, it "*loses charge*".

Gain of charge by BN condenser = Change in capacitance × p.d.

$$= \Delta C (V_o - V_b)$$

"*Gain of electrical energy*" by BN condenser

= charge × potential difference

$$= \Delta C (V_o - V_b). (V_o - V_b)$$

$$= \Delta C (V_o - V_b)^2$$

Similarly, the "*loss of electrical energy*" by AN condenser

$$= \Delta C (V_o - V_a)^2$$

Therefore, the "*net gain of energy*".

$$= \Delta C\,[(V_o - V_b)^2 - (V_o - V_a)^2]$$
$$= \Delta C\,[\{V_o - V_b + V_o - V_a\}\,\{V_o - V_b - V_o + V_a\}]$$
$$= \Delta C\,[(2V_o - (V_a + V_b)]\,(V_a - V_b)$$
$$= 2\Delta C\left[V_o - \frac{V_a + V_b}{2}\right](V_a - V_b) \qquad ...(6)$$

This is to be supplied by the source of electrical energy which is responsible for maintaining the difference of potential $(V_a - V_b)$.

The energy gain given by Eq. (6) is used in two parts:

1. In increasing the electrostatics potential energy of the system, and
2. In doing mechanical work in twisting the suspension fibre when the needle is deflected.

The increases in the electrostatics potential energy of the BN condenser

$$= \frac{1}{2}\Delta C\,(V_o - V_b)^2$$

and the loss of potential energy of the AN condenser is

$$\frac{1}{2}\Delta C\,(V_o - V_b)^2$$

Therefore, the net gain of potential energy,

$$= \frac{1}{2}\Delta C\,[V_o - V_b)^2 - (V_o - V_a)^2]$$
$$= \frac{1}{2}\Delta C\,[2V_o - (V_a + V_b)]\,(V_a - V_b)$$
$$= \Delta C\left[V_o - \frac{V_a + V_b}{2}\right](V_a - V_b) \qquad ...(7)$$

Comparison of equations (6) and 7 shows that "half" of the energy supplied by the source is used in "*increasing*" the "*electrical potential"* energy of the system. Hence the "*remaining half*" of the energy supplied by the source is used in "*twisting the suspension fibre*" during the deflection of the needle. Hence the energy available for twisting the suspension fibre is

$$= \Delta C \left[V_o - \frac{V_a + V_b}{2} \right] (V_a - V_b)$$

Putting the value of ΔC from,

Energy spent (or work done) in twisting the fibre

$$= \frac{2\varepsilon_o r^2 \theta}{d} \left[Vo - \frac{V_a + V_b}{2} \right] (V_a - V_b)$$

But the work done = twisting couple × angle of twist

Therefore, twisting or "*deflecting couple"*

$$= \frac{2\varepsilon_o r^2 \theta}{d} \left[Vo - \frac{V_a + V_b}{2} \right] (V_a - V_b)$$

If τ is the restoring couple per unit twist, then the couple for the deflection θ is

$$\tau\theta = \frac{2\varepsilon_o r^2}{d} \left[V_o - \frac{V_a + V_b}{2} \right] (V_a - V_b)$$

or $$\theta = \frac{2\varepsilon_o r^2}{d\tau} \left[V_o - \frac{V_a + V_b}{2} \right] (V_a - V_b) \quad ...(8)$$

Thus, the deflection θ is proportional to $\frac{2\varepsilon_o r^2}{d\tau}$ which is constant, to the potential deference $(V_a - V_b)$ between AA and BB and to the average difference of potential between the needle and quadrant $\left(\text{i.e.} V_o - \frac{V_a - V_b}{2} \right)$.

The quadrant electrometer can be used in two ways:

(1) Electrostatic use, and

(2) Idiostatic use

HETEROSTATIC USE

In this case the needle is charged to a very high potential as compared to the potentials Va and Vb, which are given to the quadrants AA and BB respectively. Thus, when

$$V_o > V_a \text{ or } V_b$$

Then $\frac{V_a + V_b}{2}$ may be neglected in Eq. (1) and hence,

$$\theta = \frac{2\varepsilon_o r^2}{d\tau} V_o (V_a - V_b)$$

or

$$\theta = K (V_a - V_b) \quad \text{...(1)}$$

where, $K \; \frac{2\varepsilon_o r^2}{d\tau} V_o$

Therefore, *the deflection in the heterostatic use is proportional to the potential difference.* If a graph is plotted between deflection and potential difference a straight line is obtained.

Limitation of This Method

Since $V_o > V_a$ or V_b, *this method is applicable for the measurement of the "small potential difference" only.* If the potentials applied to two pairs of quadrants are reversed the deflection will also be reversed, *hence in this use electrometer can measure only the "direct voltage"*

IDIOSTATIC USE

In the case, the needle and one pair of quadrants, say AA are connected together, hence,

$V_o = V_a$ and Eq. (1) reduces to

$$\theta = \frac{\varepsilon_o r^2}{d\tau} (V_a - V_b)^2$$

$$= K' (V_a - V_b)^2 \qquad ...(2)$$

where, $K' = \dfrac{\varepsilon_o r^2}{d\tau}$

Thus, *the deflection in the idiostatic use is proportional to the "square of the potential difference"*. Hence, *this method is used for determining " alternating potential difference"* and *"large DC potential"* when the meedle and quadrants AA are connected to the higher terminal. The variation of deflection with p.d.

Relative Merits of Heterostatic and Idiostatic Uses

We have, $\dfrac{K}{K'} = \dfrac{2\varepsilon_o r^2}{d\tau} V_o \Big/ \dfrac{\varepsilon_o r^2}{d\tau}$

$$= 2V_o \qquad ...(3)$$

Hence K > K'. Thus, *the electrometer when used electrostatically is more accurate than when used idiostatically because in the heterostatic use, we have comparatively larger deflection for the same difference of potential.* But in idiostatic use the deflection is proportional to the square of the potential, therefore, it is possible to measure alternating potential also. Again for a given deflection, the value $V_a - V_b$ is much greater on the idiostatic use than in the heterostatic use, since K' < K, hence the *range of the instrument is much more in the idiostatic use, although the accuracy is considerably less* in this case than to heterostatic use.

Uses of a Quadrant Electrometer

A quadrant electrometer can be used for the following purposes:

1. To measure small direct potential difference (or voltage). For details, see heterostatic use discussed above.
2. To measure alternating potential difference. For detail see idiostatic use discussed above.
3. Measurement of capacity and comparison of two capacities
4. Measurement of capacity current.
5. Measurement of dielectric constant.

6. Comparison of potentials.

Important cases are discussed here separately as follows:

MEASUREMENT OF CAPACITY

If one pair of the quadrants is earthed and the other is connected to a potential V, then the deflection of needle given by equations (1) *i.e.,*

$$\theta = K (V_a - V_b)$$

here, $V_b = 0$ and $V_a = V$

hence, $\theta = KV$ ...(1)

This equation is used for the measurement of the capacity of a quadrant electrometer and for the comparison of two capacities.

(a) Capacity of a Quadrant Electrometer

The circuit arrangement. Let C be the capacity of the electrometer, one pair of quadrants of the electrometer is charged to a constant V_1 with a standard cell S by depressing the key K' and the resulting deflection θ_1 is noted. Then according to equation (1), we have, $\theta_1 = KV_1$

Now K' is released and the standard condenser C_1 is put in circuit by closing the key K_1. The condenser C_1 shares the charge of the electrometer and the common potential becomes V_2 which gives the deflection θ_2 given by,

$$\theta_2 = K\Delta_2$$

then the total charge on the electrometer Q is given by,

$$Q = CV_1 = (C + C_1) V_2$$

or $$\frac{C + C_1}{C} = \frac{V_1}{V_2} = \frac{\theta_1}{\theta_2}, \left(\text{Since} \frac{KV_1}{KV_2} = \frac{\theta_1}{\theta_2}\right)$$

or $$\frac{C_1}{C} = \frac{\theta_1 - \theta_2}{\theta_2}$$

or $$C = C_1 \frac{\theta_1}{\theta_1 - \theta_2} \quad \text{...(2)}$$

Thus C can be calculated. A guard-ring capacitor is preferred as a standard condenser C_1 because its capacity can be calculated from its geometrical dimensions, as,

$$C_1 = \frac{\varepsilon_0 A}{d}$$

(b) Comparison of Two Capacities

In this case, first the condenser C1 put in the circuit by closing the key K_1 and then the key K' is depressed so that both the capacity C_1 and the electrometer are charged to a common potential V. Let θ be the corresponding deflection in electrometer. Further, the key K_1 is opened and the condenser C_2 is put in the circuit by closing the key K_2, C_2 shares some of the charge acquired by C_1 and electrometer as shown in Fig. 3.6. Consequently the potential falls to V' and the corresponding deflection becomes q', then the charge,

$$Q = (C_1 + C)\,V = (C_1 + C_2 + C)\,V'$$

or
$$\frac{C_1 + C_2 + C}{C_1 + C} = \frac{V}{V'}$$

But V Kθ and V' = Kθ'

hence,
$$\frac{C_1 + C_2 + C}{C_1 + C} = \frac{\theta}{\theta'}$$

or
$$\frac{C_2}{C_1 + C} = \frac{\theta - \theta'}{\theta} \qquad ...(3)$$

Since C, the capacity of the electrometer is known form equation (2), the ratio C_2/C_1 becomes known, however, if C is very small in comparison to C_1 and C_2, we can neglect C in equation (3),

Then,
$$\frac{C_2}{C_1} = \frac{\theta - \theta'}{\theta'} \qquad ...(4)$$

Determination of Dielectric Constant

As dielectric constant K is defined as,

$$K = \frac{\text{Capacity of condenser with the subs tan ce as dielectric}}{\text{Capacity of the condenser with air as dielectric}}$$

Hence the methods of determination of capacity of capacity and dielectric constant are same. Any method used for comparison of two capacities can be used for determining dielectric constant.

MEASUREMENT OF FEEBLE (IONISATION) CURRENTS

A quardant electrometer can be used to measure "very feeble ionisation current." The ionisation current is produced by exposing a chamber containing gas, by some ionising agent, e.g. X-rays, ultraviolet or by radiations from radioactive materials.

As ionisation chamber, consisting of a metallic box containing two metal plates A and B, insulated and separated from each other. The plate A is charged to a high positive potential and to one pair of quadrants. The other plate B is connected to one pair of quadrants of the electrometer, which is earthed. The needle of the electrometer is given a high positive potential.

When the gas in the chamber is ionised, the needle of the electrometer is deflected because due to ionisation the negative ions are produced and they move toward the electrode A. *The movement of the ions constitutes ionisation current. The rate of change of deflection is proportional to the magnitude of the ionisation current.*

Let C be the capacity of the electrometer together with the electrode A to which it is connected, and if V be the potential at any instant t due to the charge Q, the ionisation current,

$$I = \text{rate of flow of charge}$$

$$= \frac{dQ}{dt} = \frac{CdV}{dt}$$

$\theta = KV$, where K is constant.

$$\text{Hence, } I = C\,\frac{V_1 - V_2}{t} \qquad \text{...(1)}$$

$$\text{or} \qquad I = \frac{1}{K}C\frac{d\theta}{dt} = K'\,\frac{Cd\theta}{dt} \qquad \text{...(2)}$$

where K' = 1/K another constant.

$\frac{d\theta}{dt}$ is the rate of change of deflection of the needle, C is determined by experiment described earlier and constant K' = 1/K, is determined from the calibration graph. Hence I can be calculated. Currents of the order of 10^{-15} amp. can be measured with a quadrant electrometer.

MEASUREMENT OF DIELECTRIC CONSTANT

(i) Hopkinson's Null Method for Solids

We can determine the dielectric constant by this method if the solid is available in the form of slab. The circuit arrangement, C_2 is a guard ring condensers which the lower plate is movable. C_1 is a variable cylindrical condenser consisting of co-axial cylindrical conductors, the inner cylinder can slide inside the outer one, hence the capacity of this condenser can be increased or decreased by increasing or decreasing the area of overlap. The battery B is earthed at its mid-point so that its terminal points are at equal and opposite potentials with respect to the earth-point.

The needle of the electrometer is charged by battery E to a high positive potential. When the two keys K_1 and K_2 are depressed, the condensers C_1 and C_2 are charged such that the upper plate of C_1 is at a positive potential and that of C_2 to an equal negative potential. Hence, if two condensers are of equal capacity, they will acquire equal and opposite charges.

Now, if the keys K_1 and K_2 are raised such that they are in contact with C_1 and C_2 and the connection with battery is broken, the charges will neutralize one another and there will be no deflection in the electrometer. This is obtained by adjusting the capacity of the variable capacitor C_1.

Let the distance between the plates A and B of the capacitor C_2 be d. The given dielectric slab of thickness t is placed on the plate B. This increases the capacity of C_2. The previous adjustment of C_1 is kept undisturbed and the palet B is moved downward until there is again no deflection in the electrometer when K_1 and K_2 are raised such that C_1 and C_2 are connected to electrometer. Let x be the distance by which the plate B is moved.

Capacity of C_2 without dielectric $= \frac{\varepsilon_0 A}{d}$ where A is the effective area of the guard ring capacitor. Capacity of C_2 after introducing the dielectric slab, when the electrometer shows no deflection,

$$= \frac{\varepsilon_0 A}{\left[d + x - t\left(1 - \frac{1}{K}\right)\right]}$$

Therefore, $\frac{\varepsilon_0 A}{d} = \frac{\varepsilon_0 A}{\left[d + x - t\left(1 - \frac{1}{K}\right)\right]}$

or $$d = d + x - t\left(1 - \frac{1}{K}\right)$$

or $$K = \frac{t}{t - x} \qquad ...(1)$$

Hence K can be calculated.

(ii) Dielectric Constant of a Liquid by Hopkinson's Method

Procedure is the same as that for solid except that the guarding condenser C_2 replaced by a double-walled cylindrical condenser in which an insulated cylinder C could hang.

This condenser is connected in place of C_2 in circuit of and C_1 is suitably adjusted, so that there is no deflection of the needle of the electrometer. Then the capacity of condenser C is equal to the capacity of the condenser C_1.

If a and be are the inner and outer radii of the cylindrical condenser C_1 and l1 is the length of overlap of inner and outer cylinders. Then the capacity of the condenser C, is equal to the capacity of C_1 without dielectric.

$$C_{\text{without dielectric}} = \frac{2\pi\varepsilon_0 l_1}{\log_e \frac{b}{a}}$$

Now the given liquid is filled in C between A and B and the position of C_1 is again adjusted for zero deflection. If l_2 is the length of overlap then capacity of the condenser C with dielectric,

$$C_{\text{with dielectric}} = \frac{2\pi\varepsilon_o l_2}{\log_e \frac{b}{a}}$$

the ratio,
$$\frac{C_{\text{with dielectric}}}{C_{\text{without dielectric}}} = \frac{l_2}{l_1} = K \qquad \text{...(2)}$$

Hence, K, *i.e.,* dielectric constant, or the relative permittivity of the liquid, is known.

(iii) Boltzmann's Method of Determining the Dielectric Constant of Gases

The circuit used by Boltzmann. Two plates P and Q forming a parallel plate condenser are enclosed in a metal chamber C which is earthed, the chamber C one be exhausted or filled with gas. The connections from the plates P and Q are insulates from the chamber C. The plate P is connected to the positive of H.T. battery of n cells each of voltage V though key K_1. The plate Q is connected to one pair of the quadrants AA and the other quadrants are earthed. The key K_2 helps to connect plate Q and quadrants AA to the earth.

The chamber C is first evacuated then the key K_1 is closed which charges the plate P to a potential nV. If the key K_2 is simultaneously closed as K_1, there will be no deflection in the electrometer because both the Paris of quadrants are earthed. Next, keys K_1 and K_2 are opened and the chamber is filled with given gas. The capacity of the condenser formed by P and Q will increase K times and consequently, the potential of P falls to nV/K.

The key K_1 is again closed and the plate P is charged to a potential nV. It receives additional charge from the source due to its increased capacity. This additional charge induces an equal and opposite charge on Q. Thus, Q is charged by the same amount as that of P. As Q is connected to electrometer, hence electrometer needle is deflected. Let this deflection be θ, then according to eq. (3),

$$\theta \propto V$$

Hence the change in potential $= nV - \dfrac{nV}{K}$

$$= nV\left(1 - \frac{1}{K}\right)$$

Therefore, $\theta \propto nV\left(1 - \dfrac{1}{K}\right)$

or $$\theta = K'\, nV\left(1 - \frac{1}{K}\right) \qquad ...(4)$$

where K' is a constant.

Now if one more cell is added to H.T. battery then the potential of P will increase to nV + V = (n + 1) V and the potential of Q will be also increased by an equal amount. This increase of potential changes the deflection of the needle say θ', then,

$$\theta' \propto V$$

or $$\theta'\, K'\, V \qquad ...(5)$$

Hence, $$\frac{\theta}{\theta'} = n\left(1 - \frac{1}{K}\right)$$

or $$1 - \frac{1}{K} = \frac{\theta}{n\theta'}$$

or $$\frac{1}{K} = 1 - \frac{\theta}{n\theta} = \frac{n\theta' - \theta}{n\theta'}$$

or $$K = \frac{n\theta'}{n\theta' - \theta}$$

Thus knowing n, θ and θ', K can be calculated.

Comparison of Potentails

Potentials can be compared either by connecting the electometer electrostatically or idiostatically and using respectively.

Electrostatic Voltmeter

The electrostatic voltmeter was first designed by Lord Kelvin and is a modified form of quadrant electrometer. It consists of only one pair of opposite quadrants AA placed vertically. A light aluminium needle N is pivoted at the centre so that it can rotate about a horizontal axis in a vertical place symmetrically between the quadrants. The needle N is insulated from the quadrants. The upper part of the needle carries a pointer which moves over the scale S of the instrument and the lower part carries a controlling weight which can be adjusted for changing the range of the instrument.

The potential difference to be measured is applied between the needle and the quadrants. The needle N rotates due to this potential difference and the deflection of the needle is proportional to the square of the applied potential difference. The scale is directly calibrated to read volts by actual comparison with the standard cell or attracted disc electrometer. Thus, the device is handy and is capable of measuring potential of several thousand volts.

Van de Graaff Generator

This was designed by Van de Graaff in 1931. It can produce high voltage of the order of 50 mega volt. The principle of the generator is as follows. It uses the *discharge action of points and charge collecting action of hollow spherical conductors.* It consists of large hollow sphere S mounted on a large pillar P of highly insulating material.

B is an endless belt of some insulating material, *i.e.,*., rubber or silk. Belt runs on motor driven pulleys P_1 and P_2 and enters the sphere through the slit S_1 and leaves the sphere through S_2. A conductor whose one end is a point A, is charged to a high potential (about 10,000 volts) is placed near the belt. This charge leaks out to the belt. This charge leaks out to the belt which becomes positively charged. The belt moves up and passes to another close point H near P_2 connected to the metallic spheres S.

This induces a negative charge on the point H and positive charge on the sphere. The negative charge on the point H is soon neutralised by the positive charge on the belt. The sphere thus gets positively charged. As the belt is revolving continuously, the charge on the sphere is accumulating contiguously. *Consequently the potential of the sphere rises considerably depending on its size and the smoothness of its surface. The bigger the sphere and the smoother its surface, the greater is the potential it can retain without leakage.*

To prevent the discharge taking place from S, the generator is completely enclosed in an earthed steel tank which is filled with air under pressure.

Electrometers

In common practice we use the *electromagnetic voltmeters* for measuring the potential difference, but they are *less accurate* because they *require some current,* though very small, *for their working.* In electrostatics phenomena, certain instruments have been devised, in which the *current is to be completely avoided* by perfect insulation which is difficult to achieve. However, the efforts have been made to reduce, the leakage of charge due to imperfect insulation, to minimum. Thus, *accuracy is increased and we measure correctly the potential difference,* charge and capacitance. *Such devices are called electrometers.* The simplest electrometer was first used by Faraday for measuring potential difference and that was gold leaf electroscope.

MEASUREMENT OF DIELECTRIC CONSTANT USING ATTRACTED DISC ELECTROMETER

If the material is given in the form of a slab of thickness, say, t which can be introduced between the plates A and B. The effective air distance between the plates decreases and arrangement is now simply the parallel plate capacitor partially filled with dielectric, its capacity,

$$C = \frac{\varepsilon_0 A}{\left[d - t\left(1 - \frac{1}{k}\right)\right]},$$

where K is the dielectric constant of the slab. Clearly, the effect of introducing of a dielectric slab of thickness *t is effectively decreasing the distance between plates A and B by an amount* $t\left(1 - \frac{1}{K}\right)$ *Hence, if after introducing the slab the separation between the plates is increased by an amount* $h = t\left(1 - \frac{1}{K}\right)$, the capacity of the capacitor reduces to its original value, *i.e.,* $C = \frac{\varepsilon_0 A}{d}$ without the dielectric slab

Hence, $$h = t\left(1 - \frac{1}{K}\right)$$

or $$\frac{h}{t} = 1 - \frac{1}{K}$$

or $$\frac{1}{K} = 1 - \frac{h}{t} = \frac{t-h}{t}$$

or $$K = \frac{t}{t-h} \qquad ...(1)$$

Therefore the value of K can be determined in terms of t and h. The value of t and h can be easily determined with the help of attached disc electrometer when the slab is introduced between the plates A and B. The potential is obtained by replacing d in Eq. (1) by d – t $\left(1 - \frac{1}{K}\right)$, *i.e.*,

$$V_a - V_b = \left[d - t\left(1 - \frac{1}{K}\right)\right]\sqrt{\frac{2F}{\varepsilon_o A}} \qquad ...(2)$$

since $F = \frac{\varepsilon_o E^2}{2}$ and $E = \frac{V_a - V_b}{d - t\left(1 - \frac{1}{K}\right)}$

Now, if the potential difference between the plates is maintained constant and the dielectric slab introduced, the distance between the plates effectively slab introduced, the distance between the plates effectively decreases and electric field E increases so consequently F (force of attracting) increases. Hence, the plate A is pulled down by the plate B by an amount $h = t\left(1 - \frac{1}{K}\right)$ with the help of micrometer screw, the force of micrometer screw, the force of attraction will decrease to its original value and the plate A will return will decrease to its original value and the plate A will return to its original position. Hence h is known from the two readings of micrometer screw.

For measuring the thickness of the slab, the plate B is moved up so that the top and the bottom surfaces of the slab touch the plates A and B respectively. The reading of the position of B is noted. Next, the slab is removed and the plate B is again moved such that B touches A and reading of B is noted again. The difference between two readings is thickness t. Therefore, K is known by Eq. (1).

SOLVED EXAMPLES

Example 1:

Two leyden jars are exactly similar in size and shape but one has glass as dielectric and the other ebonite. The glass jar is charged by when the charge is shared between the two jars the potential falls to 0.6 of its original value. If dielectric constant of ebonite is 2, find that of glass.

Solution:

Using

$$\frac{C_1}{C_2} = \frac{\theta'}{\theta - \theta'} \qquad \text{(but } \theta = KV\text{)}$$

hence,

$$\frac{C_1}{C_2} = \frac{V'}{V - V'}$$

Let V be the original potential, the,

$$V' = \frac{6}{10}V$$

Therefore,

$$\frac{C_1}{C_2} = \frac{6V/10}{V - \frac{6V}{10}} = \frac{3}{2}$$

Since the capacitors are exactly similar their capacitance without dielectric will be same. Hence,

$$C_1 = K_1\, C_o$$

where K_1 is dielectric constant of glass

and $C_2 = K_2C_0$

where K_2 is dielectric constant of ebonite.

Then,

$$\frac{C_1}{C_2} = \frac{K_1}{K_2} = \frac{3}{2}$$

or $$K_1 = \frac{3}{2}K_2$$

$$K_{glass} = 3 \text{ (as } K_2 = 2)$$

Example 2:

When the air inside an electroscope of 20 cm capacitance is ionised by a beam of X-rays, it is found that the potential of the gold leaf changes from 154 to 100 volts in one minute. Find the current (1 farad = 9×10^{11} cm).

Solution:

$$C = 20 \text{ cm} = \frac{20}{9\times10^{11}} \text{ farad}$$

$$\text{and } I = \frac{CdV}{dt} = \frac{20}{9\times10^{11}}\times\frac{54}{60} = 2 \times 10^{-11} \text{ amp.}$$

Example 3:

Find the mass required to give a force equal to the force of attraction between the plates of attracted disc electrometer when they are 4 mm apart, having area 10 sq. cm and at a potential difference of 1000 volts.

Solution:

$$V_2 - V_1 = 1{,}000 \text{ volts}$$

and $d_1 - d_2 = 4 \times 10^{-3}$ m

and $A = 10 \times 10^{-4}$ m^2

Using Eq. (1), *i.e.*,

$$V_2 - V_1 = d_2 - d_1 \sqrt{\frac{2mg}{\varepsilon_o A}}$$

on putting the values,

$$1000 = 4 \times 10^{-3} \sqrt{\frac{2 \times 9.8 \times m}{8.85 \times 10^{-12} \times 10^{-3}}}$$

or $m = 2.8 \times 10^{-5}$ kg

Example 4:

Two large metal plates are fixed horizontally at a distance 0.5 cm from each other. What potential in volts should be applied between the plates, if a droplet of oil of mass 1.5 × 10^{-11} gms and carrying a charge 4.9 × 10^{-10} e.s.u. is to be held at rest between the plates.

Solution:

Let the potential to be applied be V volts, then electric field,

$$E = \frac{V}{0.5 \times 10^{-2}} \text{ Volts/m}$$

Upward force on the droplet

$$q\,E = \frac{V}{0.5 \times 10^{-2}} \times \frac{4.9 \times 10^{-10}}{3 \times 10^{9}} \text{ Newtons}$$

and this is balanced by weight, mg,

$$= 1.5 \times 10^{-14} \times 9.8 \text{ Newtons}$$

Hence, $\frac{V}{0.5 \times 10^{-2}} \cdot \frac{4.9 \times 10^{-10}}{3 \times 10^{9}} = 1.5 \times 10^{-14} \times 9.8$

or $V = 4500$ volts

Example 5:

An electrometer is charged by connecting it momentarily to a battery and a deflection of 100 division is observed. Then a condenser of 50 μF capacitance is connected in parallel to the electrometer and deflection falls to 75 divisions. Calculate the capacity of the electrometer.

Solution:

Let C be the capacity of the electrometer and C_1 is the capacity of the condenser. In the first case,

$$Q = CV$$

and in second case,

$$Q = (C + C_1)\ V_1$$

or $$\frac{V}{V_1} = \frac{C + C_1}{C}$$

But from equation (1), *i.e.*, $\theta = KV$

$$\frac{V}{V_1} = \frac{\theta}{\theta_1} = \frac{100}{75}$$

therefore,

$$\frac{C + c_1}{C} = \frac{4}{3}$$

or $$\frac{C_1}{C} = \frac{1}{3}$$

or $$C = 3C_1 = 3 \times 50\ \mu F$$
$$= 150\ \mu F$$

EXERCISES

1. Two brass plates are arranged horizontally 2 cm apart and the lower plate is earthed. The plates are charged to a potential difference of 400 volts. A drop of oil carrying a charge 4.8×10^{-10} esy is at rest between the plates. Calculate the mass of drop.
2. A small current flows through a resistance of 10^{10} ohms, the ends of which are connected to opposite quadrants of an electrometer. The deflection is 120 scale divisions. When a denial cell (emf 1.04 volts) is connected across these quadrants, the deflection is 54 scale divisions. Calculate the current.
3. Give the principle and theory of the attracted disc electrometer.
4. Give the theory of quadrant electrometer. Explain clearly the heterostatic and idiostatic arrangements and discuss the relative advantages.

5. What is Van de Graaff generator? Discuss its principle and working.
6. Calculate the force of attraction between lower and upper discs of an attracted disc electrometer when a potential difference of 1000 volts is applied between them. Given that they are 0.5 cm apart and area 10 sq. cm.
7. Two circular metal plates each of radius 2 cm are separated by a distance of 2mm in air. Calculate the potential difference in volts between them when the force of attraction between them is equal to one gm weight.
8. Give the theory of quadrant electrometer. How will you use it to compare two capacities?

3

Dielectric Susceptibility and Permittivity

INTRODUCTION

In most cases (in the so-called linear dielectrics) the magnitude of polarisation is directly proportional to the intensity or the electric field at a given point of a dielectric *i.e.,* we can have,

$$P \propto E$$

or $$\vec{P} = \epsilon_o \chi_{er} \vec{E} \quad \text{...(1)}$$

Substituting the dimensions in equation (1), as,

$$P = \frac{\text{couls}}{\text{meter}^2}, \; E = \frac{\text{Newtons}}{\text{Coul}},$$

$$\epsilon_o = \frac{\text{couls}^2}{\text{Newton / meter}^2} \text{m0}$$

We see that χ_{er} is a *dimensionless parameter of dielectric called relative dielectric susceptibility.* The total proportionality factor $\chi_{er} \epsilon_o = \chi e$ has the dimensions Farad/meter is known as absolute *susceptibility or dielectric susceptibility i.e.,*

$$\chi_{er} = \frac{\chi_e}{\epsilon_o} \quad \text{...(2)}$$

$$\left[\chi_{er} \epsilon_o = \frac{P}{E} = \text{coul}^2/\text{Newton. meter}^2 \right.$$

$$\text{since volt} = \frac{\text{Newton meter}}{\text{Coul}} \text{ and Farad} = \text{coul/volt}$$

$$\text{Farad} = \frac{\text{coul}^2}{\text{Newton . meter}}$$

hence, $\chi er \in_o = \left.\frac{\text{Farad}}{\text{meter}}\right]$

The polarisation vector $\vec{P}$ is also related to $\vec{E}$ *i.e.,*

$$\vec{P} = \in_o (K - 1)\vec{E}$$

We can write

$$\chi er = K - 1$$

The dielectric susceptibility is measure of the capability of the dielectric to take polarisation, by equation (2)

$$K = 1 + \chi_{er} \quad ...(3)$$

Let us now rewrite equation

$$\vec{D} = K \in_o \vec{E} = \in \vec{E} \quad ...(4)$$

where $\in$ is the permittivity of the medium,

$$\in = K \in_o$$

or $$K = \frac{\in}{\in_o} \quad ...(5)$$

K is relative permittivity or dielectric constant (sometimes relative permittivity is denoted by $\in_r$) *when a dielectric material is placed in electric field, the distribution of field changes to a degree depending upon relative permittivity.* The χ_{er} is given

$$\chi_{er} = \frac{P}{\in_o E}$$

$$= \frac{\frac{Q'}{A}}{\frac{\in_o Q}{K\in_o A}}$$

(Since $E \frac{Q}{K\in_o A}$ see equation (4.21) and $P = Q/A$ by definition given in equation (4.27))

$$= K\frac{Q'/A}{Q/A}$$

$$= \frac{\frac{\in Q'}{A}}{\frac{\in_0 Q}{A}} = \frac{\text{Charge density in dielectric}}{\text{Charge density in free space}} \qquad ...(6)$$

Now from equation (2),

$$K = 1 + \chi er$$

This relation clearly shows that the relative permittivity of any matter is larger than unity. The value of χ_{er} of any matter is positive so that K should be larger than one and for vacuum $\chi_{er} = 0$ and therefore $K = 1$.

In the case of *non-linear dielectric including ferroelectrics to be discussed later on, there is no proportionality between vectors* $\vec{P}$ and $\vec{E}$ *as discussed here.*

Potential and Field due to a Polarised Dielectric

When a dielectric is polarised the polarisation charge appears.

The polarisation vector $\vec{P}$ is defined by the equation,

$$\vec{P} = \vec{D} - \in_0 \vec{E}$$

Let us take the divergence of this equation. We have,

$$\vec{\nabla}.\vec{P} = \vec{\nabla}.\vec{D} - \in_0 \vec{\nabla}.\vec{E}$$

or
$$\in_0 \vec{\nabla}.\vec{E} = \vec{\nabla}.\vec{D} - \vec{\nabla}.\vec{P}$$

or
$$\vec{\nabla}.\vec{E} = \frac{1}{\in_0}(\rho f - \vec{\nabla}.\vec{P}), \text{ since } \vec{\nabla}.\vec{D} = \rho f \text{ by eq.} \qquad ...(7)$$

As we know that electrostatic potential due to charge distribution ρ is given by,

$$V = \frac{1}{4\pi \in_0} \int_V \frac{\rho}{r} dv',$$

where r = distance from element containing charge to point at which V is to be calculated, and ρ = true charge volume density (couls/meter^3)

When dielectric is polarised the polarisation charge appears and volume charge density due to polarisation is given by equation (7), *i.e.*,

$$\rho_p = -\vec{\nabla}'\vec{P}$$

Here prime over ∇ *is put to avoid the ambiguity with source point and point of observation because divergence is w.r.t. source point i.e., charge distribution.*

Thus, the resultant volume charge density is

$$\rho - \vec{\nabla}'.\vec{P}$$

Therefore, the total potential due to a polarised dielectric is given by,

$$V_T = \frac{1}{4\pi \epsilon_o} \int_v \frac{\rho - \vec{\nabla}'.\vec{P}}{r} dv'$$

$$= \frac{1}{4\pi \epsilon_o} \int_v \frac{\rho dv'}{r} - \frac{1}{4\pi \epsilon_o} \int_v \frac{\vec{\nabla}'.\vec{P}}{r} dv' \quad \text{...(8)}$$

Hence, the *potential* V_p *due to polarisation distribution is given by,*

$$V_p = -\frac{1}{4\pi \epsilon_o} \int_v \frac{(\vec{\nabla}'.\vec{P})}{r} dv' \quad \text{...(9)}$$

and the potential due to true charge

$$= \frac{1}{4\pi \epsilon_o} \int \frac{\rho}{r} dv' \quad \text{...(10)}$$

where, P = polarisation (couls/meter^2)

r = distance from element containing charge or polarisation to point at which V_T is to be calculated.

The volume integration is taken over all regions containing charge or polarisation.

DIPOLE APPROXIMATION TO POLARISED DIELECTRICS

The potential of a dipole is given by,

$$V(r) = \frac{1}{4\pi \epsilon_o} \frac{\vec{p}.\hat{r}}{r^2}$$

or $$V(r) = \frac{1}{4\pi \in_0}\left[-\vec{p}.\vec{\nabla}\left(\frac{1}{r}\right)\right], \left[\text{since } \nabla\left(\frac{1}{r}\right) = -\frac{1}{r^2}\right]$$

where $\vec{p}$ is the dipole moment.

If there are n dipoles per unit volume, then potential due to unit volume,

$$V(r) = -\frac{1}{4\pi \in_0} n\vec{p}.\vec{\nabla}\left(\frac{1}{r}\right)$$

$$= -\frac{1}{4\pi \in_0} \vec{p}.\vec{\nabla}\left(\frac{1}{r}\right) \quad ...(1)$$

where $\vec{p} = n\vec{p}$ *is dipole moment per unit volume. Let us compare equation (1),* then we can *interpret the integrand potential produced by fictitious polarisation by the formation of dipoles over the volume dv'. Thus, the potential V_p can then be regarded as the total potential produced by all these dipoles spread through the volume of dielectric. This means that the polarised dielectric is equivalent to an assembly of dipoles in an electric field.*

Now divergence operation (*i.e.*, $\nabla.\vec{p}$) is with respect to the charge distribution *i.e.*, source point where as the potential is determined at any external point say point of observation. Therefore, *we introduce the operator ∇' operating on the source coordinates.* The prime on the ∇ is only necessary in integral expressions which relate a field quantity to an integral over a source quantity.

Let (x, y, z) and (x', y', z') be the coordinates of the observation and source points respectively. And the scalar U is,

$$U = U(r)$$

where, $$r = [(x - x')^2 + (y - y')^2 + (z - z')^2]^{1/2}$$

then, $$\nabla U = \frac{\delta U}{\delta x}\hat{i} + \frac{\delta U}{\delta y}\hat{j} + \frac{\delta U}{\delta z}\hat{k} \quad ...(2)$$

and $$\nabla' U = \frac{\delta U}{\delta x'}\hat{i} + \frac{\delta U}{\delta y'}\hat{j} + \frac{\delta U}{\delta z'}\hat{k} \quad ...(3)$$

But, $\frac{\delta U}{\delta x} = \frac{\delta U}{\delta r}.\frac{\delta r}{\delta x}$

and, $\frac{\delta U}{\delta x'} = \frac{\delta U}{\delta r}.\frac{\delta r}{\delta x'}$

But, $\frac{\delta r}{\delta x} = \frac{1}{2}[(x - x')^2 + (y - y')^2 + (z - z')^2]^{-1/2}. 2(x - x')$

$$= \frac{x - x'}{r}$$

and, $\frac{\delta r}{\delta x'} = \frac{1}{2}[(x - x')^2 + (y-y')^2+(z - z')^2]^{-1/2} \times (-2)(x-x')$

$$= -\frac{x - x'}{r}$$

therefore, $\frac{\delta r}{\delta x'} = -\frac{\delta r}{\delta x}$

Hence, $\frac{\delta U}{\delta x'} = -\frac{\delta U}{\delta x}$

Similarly, $\frac{\delta U}{\delta y'} = -\frac{\delta U}{\delta y}$ and $\frac{\delta U}{\delta z'} = -\frac{\delta U}{\delta z}$.

putting these values in equation (3), we have,

$$\nabla U(r) = -\nabla' U(r) \quad ...(4)$$

Therefore the ∇ *in equation (1) which is with respect to the observation point using equation (4.108), becomes,*

$$V(r) = \frac{1}{4\pi \in_o}\vec{P}.\vec{\nabla}'\left(\frac{1}{r}\right)$$

Now $\vec{\nabla}'$ is with respect to source point. *The above relation can be changed into a form that is physically more meaningful by means of Gauss' divergence theorem* and the vector relation.

$$\vec{\nabla}^1.\left(\frac{\vec{P}}{r}\right) = \frac{1}{r}\vec{\nabla}'\vec{P} + \vec{P}.\vec{\nabla}'\left(\frac{1}{r}\right)$$

Substituting $\vec{\nabla}\left(\frac{1}{r}\right)$ in the expression for V(r), we get,

$$V(r) = \frac{1}{4\pi \in_o}\left[\nabla^{\bar{1}} \cdot \left(\frac{\vec{P}}{r}\right) - \frac{1}{r}\vec{\nabla}'.\vec{P}\right]$$

The potential due to the volume distribution of dipoles is found by integrating the expression over the volume which is potential due to polarised dielectrics,

$$V_p = \frac{1}{4\pi \in_0}\int\left[\vec{\nabla}'.\left(\frac{\vec{P}}{r}\right) - \frac{1}{r}\vec{\nabla}'.\vec{P}\right]dv'$$

or
$$V_p = \frac{1}{4\pi \in_o}\left[\int\vec{\nabla}'.\left(\frac{\vec{P}}{r}\right)dv' - \int\frac{\vec{\nabla}'.\vec{P}}{r}dv'\right]$$

The first term by Gauss' divergence theorem,

$$\int_V \vec{\nabla}'\left(\frac{P}{r}\right)dv = \int_S \frac{\vec{P}.\hat{n}}{r}dS'$$

Thus,
$$V_p = \frac{1}{4\pi \in_o}\int_S \frac{\vec{P}.\hat{n}}{r}dS' - \frac{1}{4\pi \in_o}\int\frac{\vec{\nabla}.'P}{r}dv'$$

We know the $\sigma_p = \vec{P}.\hat{n}$ and $\rho_p = -\vec{\nabla}.'P$

This expression can be interpreted as follows. *The first term, a surface integral, is a potential equivalent to that of a surface charged density while the second term is a potential equivalent to that of a volume charge density. Thus,*

$$V_p = \frac{1}{4\pi \in_o}\int_S \frac{\sigma_p dS'}{r} + \int\frac{\rho_p dv'}{r}, \quad ...(5)$$

Therefore, the total potential produced by a charge distribution ρ in the presence of dielectric or the potential due to a polarised dielectrics given by,

$$V_T = \frac{1}{4\pi \in_o}\int\frac{\rho}{r}dv' + \frac{1}{4\pi \in_o}\int\frac{\rho_p dv'}{r} + \frac{1}{4\pi \in_o}\int_S\frac{\sigma_p dv'}{r} \quad ...(6)$$

The electric field E produced by ρ in the presence of a dielectric can be expressed by,

$$\vec{E} = -\vec{\nabla} Vr$$

$$= -\frac{\delta Vr}{\delta r}$$

$$\vec{E} = \frac{1}{4\pi \in_o} \int \frac{\rho dv'}{r^2} \hat{r} + \int \frac{\rho_p dv'}{r^2} \hat{r} + \int \frac{\sigma_p dS'}{r^2} \hat{r} \quad ...(7)$$

where $\hat{r}$ is unit vector directed from the source point to point of observation. The electrostatics field lines begin and end on electric charges. According to Eq. (7), the sources of the electrostatics field $\vec{E}$ in the presence of dielectrics are not only the real charges ρ, but also the fictitious, polarisation charges ρ_p and σ_p. Therefore, the field lines of $\vec{E}$ begin and end not only at the points where ρ is present, but also no dielectric surfaces (σ_p) as well as at the points within the dielectrics (r_p). The field lines of $\vec{D}$ on the other hand, begin and end only on the real charge ρ. This follows from the divergence law $\nabla . \vec{D} = \rho$, which shows that $\vec{D}$ is always associated with the real charges only, so that real charges are sources of $\vec{D}$, therefore an $\vec{E}$ and a $\vec{D}$ field may have entirely different structure when dielectrics are present.

BOUNDARY CONDITIONS AT THE SURFACE SEPARATING TWO SUBSTANCES

Let us examine, how the electric field changes at the boundary between two different media. The electric field change that occurs in going from one medium to another is determined by the two basic ideas in electrostatics, namely,

(i) the first is Gauss's law. *i.e.*, $\oint_S \vec{D}.d\vec{S} = Q$ and

(ii) the second is that an electrostatics field is a conservative field in other words, on work is done in transporting a charge around a closed path in an electrostatics field, *i.e.*,

$$\oint \vec{E}.\, d\vec{l} = 0$$

We shall apply Gauss's law to the cylindrical surface of height h and base area Δs as. The cylindrical box is so constructed that it lies half in each medium. Let D_{1n} be the average normal component of displacement vector $\vec{D}$ to the bottom of the box in medium 1 and D_{2n} the average normal component of displacement vector $\vec{D}$ to the face of the box in medium 2. D_{1n} *is an inward normal.* By making the height of the cylinder h approaching zero, the contribution of the curved surface to the flux is taken as zero. Thus by Gauss's law, the total flux,

$$D_{2n}\,\Delta S - D_{1n}\,\Delta S = Q$$

where Q is total enclosed by the surface.

The second term of left hand side of this equation is negative because D_{1n} *and* ΔS *are oppositely directed.* We can write,

$$D_{2n} - D_{1n} = \frac{Q}{\Delta S} = \sigma \qquad ...(1)$$

where σ is the charge per unit area on the boundary of the two substances. *According to equation (1) the normal component of the displacement vector* $\vec{D}$ *changes at a charged boundary between two dielectrics by an amount equal to the surface charge density.*

If the boundary is free from charge $\sigma = 0$, than equation (1) reduces,

$$D_{2n} - D_{1n} \qquad ..(2)$$

Thus, the normal component of displacement vector is continuous across the charge free boundary between two dielectrics.

We shall now use the idea that the electrostatics field is conservative, the integral of $\vec{E}$ around a closed path,

$$\int_{ABCD} \vec{E}.\vec{dl} = \int_{AB} \vec{E}.\vec{dl} + \int_{BC} \vec{E}.\vec{dl} + \int_{CD} \vec{E}.\vec{dl} + \int_{DA} \vec{E}.\vec{dl} = 0$$

By making the path length BC very small approaching zero, the work along the segments BC and DA of the path normal to the boundary is zero even though a finite electric field may exist normal to the boundary. Therefore, the line integral of $\vec{E}$ around ABCD rectangle is,

$$E_{1t}\,\Delta x - E_{2t}\,\Delta x = 0,\ AB = CD = \Delta x$$

or

$$E_{1t} = E_{2t} \qquad ...(3)$$

Thus the tangential components of the electric field are the same on both sides of a boundary between two dielectrics. In other words, the *tangential electric field is continuous across such a boundary.*

BOUNDARY CONDITIONS AT THE SURFACE OF A CHARGED CONDUCTOR

If one of the substances (medium or dielectric) considered above is a conductor, for example, if medium 1 is a conductor, then the electric field inside the conductor is zero, consequently, since there are no permanent dipoles in the conductors, the polarisation $\vec{P}$ must be zero, then according to relation ($\vec{D} = \epsilon\, \vec{E} + \vec{P}$) $\vec{D}$ is also zero,

$$D_{1n} = 0$$

Thus reduces,

$$D_{2n} = \sigma \qquad ...(1)$$

This expresses the important result that the normal component of the displacement just out side the conductor is equal to the surface charge density on the conductor. As medium 1 is a conductor ($\sigma_1 \neq 0$) the field E_{1t} in medium 1 must be zero under static condition.

$$E_{2t} = 0 \qquad ...(2)$$

That is, the tangential component of the electric field just outside a conductor (a dielectric conductor boundary) must be zero.

Flux Tubes

Maxwell and Faraday extended the idea of *flux lines to the flux tube which* have a definite quantitative significance. Let us consider the field intensity at a constant radius from a point charge, the intensity is same for all points on the circumference of this radius. So that the $\vec{E}$ *and the flux density* $\vec{D}$ *are continuous functions of position around the point charge, and both are constant for a fixed radius.*

Thus, if no flux lines pass through a certain area surrounding the point charge this means flux density is zero there which is not true because $\vec{D}$ and $\vec{E}$ have constant value for all points at equal distances from the charge. It may be that some lines would pass through the area if a larger number of flux lines had been assumed. This difficulty can be avoided by simple extension of the concept of flux lines to flux tubes.

A flux tube is defined as an imaginary tube with walls that are everywhere parallel to $\vec{D}$ *and with a constant electric flux over any cross-section. The requirement that the flux over any cross-section be a constant actually is a consequence of the fact that* $\vec{D}$ *is parallel to the sides of the tube and, therefore, that the flux over the side walls is zero. The cross section of the tube may be of any shape. Conveniently tubes of triangular, squares, or hexagonal cross-section are preferred. In single flux tube of square cross-section is shown which originates at +Q and ends on –Q.*

The magnitude of $\vec{D}$ *varies along the tube but the integral of the normal component of* $\vec{D}$ *over any cross-section of the tube is constant.*

In *the flux density* $\vec{D}$ *is indicated at three points along the tube, where the magnitude of* $\vec{D}$ *is large, the cross-sectional area is small and vice versa. The integral of* $\vec{D}$ *over each cross-sectional area is constant.* All the space can be divided up into tubes of equal flux originating on positive charge.

Number of Tubes

In rationalised RMKS system, we have,

$$\oint_S \vec{D}.\vec{dS} = Q$$

Thus *each unit charge gives rise to a unit tube* (*in place of* 4π *tubes in unrationalised system*).

FARADAY TUBES (UNIT TUBES OF INDUCTION)

In RMKS, each unit charge gives rise to a unit tube irrespective of the surrounding medium. The tubes are then referred as unit tubes of induction. The number or unit tubes of induction due to a charge q placed in vacuum per unit area

$$= \frac{q}{4\pi r^2}$$

$$= \epsilon_0 \frac{q}{4\pi\epsilon_0 r^2} = \epsilon_0 \vec{E} \qquad ...(1)$$

i.e., flux density $\vec{D}$.

Maxwell Tubes

However, Maxwell assumed that $1/\epsilon$ unit tubes originate from each unit charge in a medium of permittivity ϵ. Thus the charge q gives rise to $q/\in$ Maxwell's tubes. If we draw a sphere of radius r surrounding the charge, then,

$$\text{No. or Maxwell's tubes per unit area} = \frac{q}{\in 4\pi r^2}$$

$$= \vec{E} \quad \text{...(2)}$$

Hence, the electric intensity at the point is the number of Maxwell unit tubes of force over a unit area surrounding the point.

Maxwell assumed that these tubes are under *longitudinal tension of* $\in_0 \frac{E^2}{2}$ Newton/meter^2 along the direction of the tube which tends to contract and the *lateral pressure of* the same value $\left(\in_0 \frac{E^2}{2} \text{ Newton / meter}^2\right)$ normal to the length, and showed that such a system of tension and pressures in the medium are in equilibrium. He explained that the *attraction between two opposite charges is due to longitudinal tension tending to shorten the tube. The repulsion between the charges could similarly is due to the pressure of the tubes upon each other at right angles to their length and which tend to push the two charges further apart.*

Suppose a tube of force. Apply Gauss's law at P and Q planes of areas dS_1 and dS_2.

$$\in_1 E_1 dS_1 = \in_2 E_2 dS_2 = 0$$

Since there is no charge enclosed by the tube where $\in_1$ and $\in_2$ are the permittivity constant at these two surfaces.

Thus, $$\in_1 E_1 dS_1 = \in_2 E_2 dS_2$$

or $$\int_S \vec{D}_1.d\vec{S}_1 = \int_S \vec{D}_2.d\vec{S}_2$$

Thus, electric displacement vectors (or flux densities) are different but flux is same.

Types of Capacitors According to Shape

The value of C given depends on the size and geometry of the plates, hence we can design the capacitors of different geometry, the more commonly used capacitors are,

1. a parallel plate capacitor
2. a cylindrical capacitor
3. a spherical capacitor

A CYLINDRICAL CAPACITOR

Such a capacitor consists of two coaxial cylinder of radius a and b and length l. Calculate capacitance of device, sectional diagram is given in Fig. 4.2. Construct a Gaussian surface shown by dotted line having radius r, by using Gauss's law we get,

$$\in_0 \oint \vec{E}.d\vec{S} = Q$$

where Q is the charge on the surface.

$$\in_0 E.2\pi rl = Q$$

$$E \frac{Q}{2\pi \in_0 rl} \text{ at P}$$

the potential difference between the plates is given by,

$$V = -\int_b^a \vec{E}.d\vec{r}$$

As the angle between $\vec{E}$ and $d\vec{r}$ is 0, since $\vec{E}$ is radial *i.e.,* along $\vec{r}$, so $\vec{E}.d\vec{r} = Edr$, Hence,

$$V = -\int_b^a E\,dr = -\int_b^a \frac{Q}{2\pi \in_0 rl} dr$$

$$= \frac{Q}{2\pi \in_0 l} \log_e \frac{b}{a}$$

Therefore, the capacitance,

$$C = \frac{Q}{V} = \frac{2\pi \in_0 l}{\log_e (b/a)} \quad \text{...(1)}$$

Like the capacitance C given in equation (1) *also depends on geometrical factors.*

SPHERICAL CAPACITORS

Let us calculate the capacitance of two concentric spheres. The potential difference between the two spheres V, given by,

$$V = -\int_b^a \vec{E}.d\vec{r}$$

$\vec{E}$ at a distance r is given by,

$$\vec{E} = \frac{Q}{4\pi \epsilon_o} \frac{\hat{r}}{r^2}$$

$\vec{E}$ is radial *i.e.,* along r, hence angle between $\vec{E}$ and $\vec{r}$ is zero. Therefore , $\vec{E}.d\vec{r}$ = Edr. Hence,

$$V = -\int_b^a \frac{Q}{4\pi \epsilon_o} \frac{dr}{r^2} = -\frac{Q}{4\pi \epsilon_o}\int_b^a \frac{dr}{r^2}$$

$$= \frac{Q}{4\pi \epsilon_o}\left(\frac{1}{a} - \frac{1}{b}\right) \text{ and } C = 4\pi \epsilon_o \frac{ab}{b-a} \qquad ...(1)$$

Here again the capacitance C depends only on their radii.

GUARD RING CAPACITOR

To overcome the edge effect in parallel plate capacitor the two rings one on right and other on left side are attached. These rings are called guard ring to ensure uniformity in field and such capacitor is called guard ring capacitor.

ENERGY STORED IN A CAPACITOR

The energy of a charged capacitor is the amount of work done in charging it. Suppose V is the potential difference between the plates of the capacitor of capacitance C. The charge Q is given by,

$$Q = CV$$

There is charge Q on one plate and — Q on the other. Let us increase the charge from Q to Q + dQ by transferring a positive charge dQ from the negative to the positive plate, working against the potential difference V. The work that has to be done is given by,

$$dW = VdQ = \frac{Q}{C}dQ$$

Therefore, to charge the capacitor starting from the uncharged state to the final charge say, Q

$$W = \int_{Q=o}^{Q=Q} \frac{Q}{C}dQ = \frac{1}{2}\frac{Q^2}{C} = \frac{1}{2}CV^2 = \frac{1}{2}QV \qquad ...(1)$$

This work done is stored up in the form of potential energy and supplies the energy necessary to drive the current when the capacitor is discharged.

For a parallel plate capacitor the C is given by,

$$C = \frac{\epsilon_o A}{d} \text{ and } V = Ed$$

Hence the energy stored is given by,

$$U = \frac{1}{2}\frac{\epsilon_o A}{d}E^2.d^2 = \frac{1}{2}\epsilon_o E^2.A.d \text{ joules} \qquad ...(2)$$

Energy stored per unit volume

$$= \frac{1}{2}\epsilon_o E^2 \text{ joules/metre}^2 \qquad ...(3)$$

(A.d is the volume between the plates). This expression for energy is same as the general formula for the energy in an electric field.

FORCE OF ATTRACTION BETWEEN CAPACITORS PLATES

Consider a simple parallel plate capacitor such as illustrated. It appears reasonable, since one plate is positively charged and the second plate is negatively charged, that there should be a force of attraction between the plates tending to pull them together.

An expression for the force can be readily obtained by the application of the *principle of virtual work.* Suppose E is intensity of electric field between the plates. The total energy in the field is given by,

$$U = \frac{1}{2}\epsilon_o E^2Ad \text{ joules}$$

Let us now calculate the change in the total energy when the capacitor plate to be separated by an additional amount Δd. *The work that would be required to effect the additional separation is equal to the change in energy in the field,* and this, must be equal to the product of the force and distance. The energy per unit volume is given by,

$$U = \frac{1}{2} \epsilon_o E^2 A d \text{ joules/metre}^3$$

The change in volume due to Δd change in separation = $A \Delta d$. Hence, the change in total energy due to change in volume

$$= \frac{1}{2} \epsilon_o E^2 A \Delta d \quad \text{...(1)}$$

The mechanical work done from the outside against the force of attraction between the plates in moving them further apart through a distance Δd will be,

$$W = \vec{F}.\Delta\vec{d} \quad \text{...(2)}$$

or

$$W = -F \Delta d \quad \text{...(3)}$$

This work done = change in energy

$$\frac{1}{2} \epsilon_o E^2 A \Delta d = \vec{F}.\Delta\vec{d}$$

or

$$\frac{1}{2} \epsilon_o (\vec{E}.\vec{E}) A \Delta d = \vec{F}.\Delta\vec{d} = F \Delta d$$

The *negative sign indicates that the force acting over the area A is normal to the plates in the direction of* E, *i.e.,* F *and* Δd *are* 180° apart. The magnitude is given.

$$F = \frac{1}{2} \epsilon_o E^2 A \quad \text{...(4)}$$

But the force per unit area is pressure, and hence the pressure is given by,

$$P \text{ (Pressure)} = \frac{F}{A} = \frac{1}{2} \epsilon_o \vec{E}\vec{E}$$

or

$$P = \frac{1}{2} \epsilon_o E^2 \text{ newton/ metre}^2 \quad \text{...(5)}$$

This expression for pressure is the same as the general formula for pressure on a conductor due to electric field $\vec{E}$ discussed.

THE DIELECTRIC CONSTANT

Now *we will discuss, insulator or dielectric materials which do not conduct electricity.* However, using a simple electroscope and two parallel plate capacitors Faraday discovered that this was not so. He constructed two identical capacitors in one of which he placed a dielectric. When both capacitors were charged to the same potential difference.

Faraday found by experiment that the charge on the capacitor with dielectric is greater than that without. *Since* q *is larger, for the same V, it follows from the relation* C = q/V *that the capacitance of a capacitor increases if dielectric is placed between the plates.* If dielectric material fills completely, the space between the plates *the ratio of the capacitance* C *with the dielectric to that without* C_0 *is called dielectric constant,* K, *which is the property of dielectric material i.e.,*

$$C/C_0 = K \qquad ...(1)$$

K is also *called relative permittivity* of the medium.

Faraday obtained the same relation Eq. (1) by placing same charge to the capacitors and measured the potential difference. If V_d is the potential difference with dielectric and V_0 without dielectric capacitor then $V_d = \frac{V_0}{K}$. So the potential difference with dielectric decreases by a factor 1/K. Hence from relation C = q/V as q is same, here again the same conclusion.

CLASSIFICATION OF DIELECTRIC ON THE BASIS OF MOLECULAR STRUCTURE (POLAR AND NON POLAR MOLECULES)

The molecules or atoms are composed of electrons and nuclei in a state of vibration about certain fixed points in space. Though the molecule or atom as a whole is neutral *i.e.,* the magnitude of charge on electrons is equal to the positive on nuclei due to protons. But the spacial arrangement of charges in a molecule may be different in various substances. All positive and negative charges of a molecule can be replaced by an equivalent single positive and single negative charge located at the centres of gravity of separate positive and separate negative charges respectively. These centres of gravity can be found with the aid of the same rules which are used to determine the position of the centres of gravity of systems of point masses in statics. These centres of gravity may either coincide or not coincide in space.

At any given instant, if the centre of gravity of the positive and negative electrical charges "do not coincide", then the molecule is "polar" and the matter composed of such molecules is also called polar. When the two centres of gravity are separated, the molecules may momentarily possess a dipole moment equal to the product of charge and distance of separation. While if the two centres of gravity "coincide." then the molecules is called a "non-polar molecule" having negligible dipole moment and the matter composed of such molecules is also called non-polar. Matter comprising different molecules both non-polar and polar should be regarded as polar matter.

Examples of Non-Polar Molecules

The rigid molecules which have symmetrical structure are non-polar and have always as negligible small dipole moment. They are

$$H_2, N_2, Cl_2, CH_4, CCl_4, C_6H_6, C_2H_2$$

and C_2H_4.

Example of Polar Molecules

Most molecules are, however, polar, there average dipole moment is finite. *If symmetry is disturbed,* the molecule becomes polar. If a hydrogen atom of methane molecule is replaced by chlorine atom, it losses symmetry and becomes polar.

A DIELECTRIC IN AN ELECTROSTATIC FIELD (POLARISATION OF DIELECTRIC)

We shall discuss this in the next section in detail, here is an *elementary atomic view. A dielectric whether polar or nonpolar will acquire dipole moment by induction when placed in an electric field.* Then these elementary dipoles are oriented along the field direction, dielectric becomes polarised without affecting the neutrality of dielectric. In other words, *by induction the one side get positive charge while other the same negative charge.* So the electric field setup by them (E') opposes the external electric field say E_0. *One important point to be noted is that induced effect is present only when the electric field is present.* The resultant field $\vec{E} = \vec{E}_0 + \vec{E}'$. The field E' due to induced charge points opposite to E_0, hence the field inside dielectric decreases same as potential. So we can write, using the relation,

$$E = Vd$$

$$\frac{V_0}{Vd} = \frac{E_0}{E} = K \qquad ...(1)$$

CAPACITY OF A PARALLEL PLATE CAPACITOR PARTIALLY FILLED WITH DIELECTRIC

A capacitor partially filled with a dielectric material of thickness t. If Q is the charge on each plate of area A. Then value of electric field E_0 due to charge Q in the free-space between plates and dielectric material given by Gauss's law

$$\epsilon_0 \oint \vec{E}.d\vec{S} = \epsilon_0\, E_0 A = Q$$

or
$$E_0 = \frac{Q}{\epsilon_0 A} \qquad ...(1)$$

From Eq. (1) the field E inside the dielectric is given by,

$$E \frac{E_0}{K} = \frac{Q}{K \epsilon_0 A} \qquad ...(2)$$

E_0 is the value of electric field in the free space (*i.e.,* without dielectric) which is dt, and E is the value of electric field inside the dielectric of thickness t. The potential difference between the two points is given by,

$$(V = V_1 - V_2 = -\int_2^1 \vec{E}.d\vec{l})$$

Hence the total potential difference between the plates A and B,

$$V_a - V_b = -\underset{\substack{\text{for}\\\text{free}\\\text{space}\\d-t}}{\int \vec{E}_0.d\vec{l}} - \underset{\substack{\text{for}\\\text{thick}\\\text{ness}\\t}}{\int \vec{E}.d\vec{l}}$$

Since the total field is $(\vec{E}_o + \vec{E})$

Therefore,

$$V_a - V_b = Et + E_o (d - t)$$

$$(\vec{E}.d\vec{l} = -Edl, \text{similarly, } \vec{E}_o.d\vec{l} = -E_o dl)$$

Substituting E_o and E from equation (1) and (2), we get,

$$V_a - V_b = \frac{Q}{A \epsilon_o}\left(d - t + \frac{t}{k}\right)$$

Therefore, from the capacitance C is given by,

$$C = \frac{Q}{V_a - V_b} = \frac{\epsilon_o A}{\left[d - t\left(1 - \frac{t}{k}\right)\right]} \quad (3)$$

DIELECTRICS AND GAUSS'S LAW

Now apply Gauss's law to a parallel plate capacitor filled with a dielectric of dielectric constant K. An usual parallel plate capacitor without dielectrics. Let charge be Q on the plates with proper sign and A be the area of each plate. E_o is the electric field without dielectric.

Therefore, the Gauss's law gives,

$$\epsilon_o \oint \vec{E}.d\vec{S} = \epsilon_o E_o A = Q$$

or $$E_o = \frac{Q}{\epsilon_o A}$$

Now place the dielectric then the net charge enclosed by Gaussian surface is Q – Q'. Therefore with dielectric the Gauss's law is,

$$\epsilon_o \oint \vec{E}.d\vec{S} = Q - Q' \quad ...(1)$$

or $$\epsilon_o EA = Q - Q'$$

or $$E = \frac{Q}{\epsilon_o A_o} - \frac{Q'}{\epsilon_o A} \quad ...(2)$$

in which – Q' is the induced charge due to polarisation. Using, we have

$$E = \frac{E_o}{k}$$

where E is field inside the dielectric substituting E_o, we have,

$$E = \frac{Q}{k \in_0 A} \qquad ...(3)$$

Substituting this value of E in Eq. (2), we have,

$$\frac{Q}{k \in_0 A} = \frac{Q}{\in_0 A} - \frac{Q'}{\in_0 A} \qquad ...(4)$$

or $$Q' = Q\left(1 - \frac{1}{k}\right) \qquad ...(5)$$

If there is no dielectric between the plates then K = 1, and Q' = Q, *i.e., there will be no induced charge.* The value of K is always greater than one (K > 1), hence Q' (induced charge) is always less than Q (free charge).

From Equation (5) the value of Q – Q' is given by,

$$Q - Q' = \frac{Q}{k}$$

Now we can write the Gauss's law with dielectric as given after substituting Q – Q' in the following form, *i.e.,*

$$\in_0 \oint \vec{E}.d\vec{S} = Q - Q'$$

or $$= \frac{Q}{k}$$

or $$\in_0 \oint K\vec{E}.d\vec{S} = Q \qquad ...(6)$$

Though we have derived equation (6) for a parallel plate capacitor, but it is applicable in all cases when the dielectrics are present.

CAPACITY OF A PARALLEL PLATE CAPACITOR WHEN COMPLETELY FILLED WITH DIELECTRIC

Represents such capacitor. In such circumstances, the value of intensity of electric field is E between the plates is given by,

$$E = \frac{Q}{k \in_0 A}$$

and thickness of the slab is same as the separation between the plates *i.e.,* d. Hence,

$$V = Ed$$

$$V = \frac{Qd}{k \in_0 A}$$

therefore, $$C = \frac{Q}{V} = \frac{k \in_0 A}{d} \text{ Farad}$$

ELECTRIC POLARISATION VECTOR P

The polarisation P is defined as the induced surface charge per unit area i.e., surface density of bound charges in dielectric,

$$P = \frac{Q'}{A} = \sigma \text{pol} \qquad ...(1)$$

The induced charge Q' appears when the dielectric is polarised that is the reason we call Q'/A as electric polarisation P.

Now multiply the numerator and denominator of Eq. (1), by d the thickness of the dielectric slab we get,

$$P = \frac{Q'd}{Ad}$$

The product Q'd is the induced electric dipole moment because by polarisation the dielectric slab develops the charge Q' on one surface and —Q' on the opposite separated by thickness of the slab, hence a dipole of dipole moment Q'd. The product A.d is the volume (v) of the slab therefore, we can write,

$$P = \frac{\text{Induced dipole moment}}{\text{Volume}} = \frac{Q'd}{v} \qquad ...(2)$$

The polarisation P is defined as the induced dipole moment per unit volume. Since the electric dipole moment is a vector the electric polarisation is also a vector and its magnitude is P.

$$\vec{P} = \frac{Q'\vec{d}}{v} \qquad ...(3)$$

Thus the magnitude P is the polarisation charge per unit area on the surface, *i.e.*, the surface density of charges (σ_{pol}) appearing at right angles to the direction of the applied field. Hence,

$$\sigma_{pol} = \frac{Q'}{A} = p \qquad ...(4)$$

The equation (4) can be generalised by considering the case when dielectric surface is not perpendicular to P.

Let the normal to the surface OO' make an angle θ with the direction of P. The component of P normal to the surface is equal to $\vec{p}.\hat{n}$, where $\hat{n}$ is a unit vector along the normal to the surface, we have,

$$\sigma \text{ pol} = \vec{p}.\hat{n} \qquad ...(5)$$

$\vec{p}$ at a Point

The value of P in equation (2) is an average for the volume Ad. To define the meaning of p at a point, it is convenient to assume that a dielectric in an electric field has *continuous distribution of infinitesimal dipoles,* that is a continuous polarisation. Let us consider a continuously polarised dielectric, *the value of* $\vec{p}$ *at a point can be defined as the net dipole moment* $\vec{p}$ *of a small volume* Δ v *divided by the volume,* with the limit taken a Δv shrinks to zero around the point, thus,

$$\vec{P} = \lim_{\Delta v \to 0} \frac{\vec{P}}{\Delta v} \qquad ...(6)$$

UNIFORM AND NON-UNIFORM POLARISATION

When isotropic and homogeneous materials (*i.e.*, the materials having constant value of K hence the constant value of dipole moment or polarisation vector $\vec{p}$) are placed in an electric field developed opposite charges of equal magnitude of the two sides of the surface because of polarisation. *Hence the net charge or average charge density is zero.* Such polarisation is called "uniform polarisation". The charge density remains the positive and, negative charges are displaced relative to each other. Their displacement does not produce any net charge inside the volume.

On the other hand, if the material is *non-isotropic, and in homogeneous,* K and $\vec{p}$ are not constant, say $\vec{p}$ is large at one place and smaller at another. This means that more charge be moved into some region than away from it. We then get a *volume density of* charge. The total charge displaced out of any volume v by the polarisation is the integral of the outward drawn normal component of P over the surface that bounds the volume. An equal charge of the opposite sign is left behind when charge crosses the surface. Let the net charge enclosed by volume be ΔQ_p. We can write,

$$\Delta Q_p = -\int_S \vec{\sigma}_p . d\vec{S}$$

Now using relation, $\sigma_{pol} = \vec{p}.\hat{n}$, see equation (1)
Hence,

$$\Delta Q_p = -\int_S \vec{P}.\hat{n} dS$$

If ρ_p is the volume density of polarised charge in the dielectric the,

$$\Delta Q_p = \int_V \rho_p \, dv$$

equating these two values, we get,

$$\int_V \rho_p \, dv = -\int_S \vec{P}.\hat{n} dS$$

using Gauss's divergence theorem, i.e,

$$\int_S \vec{P}.\hat{n} \, dS = \int_V \text{div}.\vec{P} dv = \int_V \vec{\nabla}.\vec{P} \, dv$$

Hence above relation reduces,

$$\int_V \rho_p \, dv = -\int_V \vec{\nabla}.\vec{p}.dv$$

or $$\vec{\nabla}.\vec{p} = -\rho_p \qquad ...(2)$$

In uniform polarisation process which occurs in isotropic materials the value of $\vec{p}$ *is constant throughout the material, hence* $\nabla\vec{p}.$ *is zero*. Now let us consider the *non-uniform polarisation.* $\vec{p}$ is not constant, it varies from point to point inside the dielectric material, than $\vec{p}$ is function of position and $\nabla.\vec{p}$ *is given by equation* (2).

THREE ELECTRIC VECTORS

Let us consider the polarisation of a dielectric slab placed between the plates of parallel plate capacitor and consider equation (1), *i.e.*,

$$\frac{Q}{K \epsilon_0 A} = \frac{Q}{\epsilon_0 A} - \frac{Q'}{\epsilon_0 A}$$

or $$\frac{Q}{A} = \epsilon_0 \left(\frac{Q}{K \epsilon_0 A} \right) + \frac{Q'}{A} \quad ...(1)$$

$$\frac{Q}{K \epsilon_0 A} = E$$

and from, $$\frac{Q'}{A} = P$$

After substitution, equation (1) reduces,

$$\frac{Q}{A} = \epsilon_0 E + P \quad ...(2)$$

Put the ratio, $\frac{Q}{A} = D$ ...(3)

We call it the electric displacement an equivalent term for flux density. As E and P on the right hand side of equation (2) are vectors, D is also a vector, hence we can write,

$$\vec{D} = \epsilon_0 \vec{E} + \vec{P}, \textit{ everywhere} \quad ...(4)$$

or $$\vec{D} = \left(\epsilon_0 + \frac{\vec{P}}{\vec{E}} \right) \vec{E} \quad ...(5)$$

The *displacement vector* is an important addition which is of *great use in Maxwell electro-magnetic equation,* to explain the displacement current. However, from equation (3), it is clear that $\vec{D}$ is associated only with free charges. The representation of $\vec{D}$, $\epsilon_0 \vec{E}$ and $\vec{P}$ for a parallel plate capacitor with dielectric.

Be careful to note the following points :

(1) $\vec{D}$ *is connected with the free charge only.* The lines of $\vec{D}$ begin and end on free charges. $\vec{D}$ is constant throughout the capacitor.

(2) $\vec{P}$ *is connected with polarisation charge only.* The line of $\vec{P}$ begin and end on polarisation charges. Clearly $\vec{P}$ is zero except inside the dielectric.

(3) $\vec{E}$ *is connected with all charges.*

(4) *In isotropic media* $\vec{P}$ and $\vec{E}$ are in the same direction. However for, *nonisotropic media* $\vec{P}$ and $\vec{E}$ *are, in general, not in the same direction.*

(5) $\vec{E}$ is reduced inside the dielectric, where there are fewer lines.

(6) Unlike the electric field

(which is the force acting on unit charge) or the polarisation $\vec{P}$ (the dipole moment per unit volume), the electric displacement $\vec{D}$ has no clear physical meaning. The only reason for introducing it is that it enables one to calculate fields in the presence of dielectrics without first having to know the distribution of polarisation charges.

Units of $\vec{D}$ and $\vec{P}$

From units for $\vec{P}$ and $\vec{D}$ are cous/meter^2 whereas $\vec{E}$ is measured in Newtons/coul.

$\vec{D}$ and $\vec{P}$ in Terms of $\vec{E}$

The vectors $\vec{D}$ and $\vec{P}$ are separately connected to $\vec{E}$. Let us start from the definition of $\vec{D}$, the magnitude of $\vec{D}$ is given by,

$$D = \frac{Q}{A} = K \in_0 \left(\frac{Q}{K \in_0 A} \right)$$

But $\dfrac{Q}{K \in_0 A} = E$, the expression for $\vec{D}$ can be written as,

$$\vec{D} = K \in_0 \vec{E} = \in \vec{E}, \text{ (as } \in = K\in_0) \qquad ...(6)$$

Hence relation $\vec{D} = \in_0 \vec{E} + \vec{P}$ can be written as,

$$\in \vec{E} = \in_0 \vec{E} + \vec{P}$$

or $$\in = \left(\in_0 + \frac{\vec{P}}{\vec{E}} \right) \qquad ...(7)$$

Similarly, we can express $\vec{P}$ in terms of $\vec{E}$ the magnitude of P is given by equation (7) *i.e.,*

$$P = \frac{Q'}{A}$$

The induced charge Q' due to polarisation is given in equation (8) *i.e.,*

$$Q' = Q \left(1 - \frac{1}{K}\right)$$

Substituting this value in the above equation, we have,

$$P = \frac{Q}{A} = \left(1 - \frac{1}{K}\right)$$

Since, $\frac{Q}{A} = D$, using eq. (9), we get,

$$\vec{P} = K \in_0 \vec{E} \left(1 - \frac{1}{K}\right)$$

or $$\vec{P} = \in_0 (K - 1) \vec{E} \qquad ...(10)$$

In vacuum K = 1, therefore the polarisation vector $\vec{P}$ is zero.

Restatement of Gauss's Law or (Divergence of $\vec{D}$)

The definition of $\vec{D}$ given by equation (9) when substituted then the simple modified form of the Gauss's law in the presence of dielectric is written as,

$$\oint_S \vec{D}.dS = Q \qquad ...(11)$$

or in words, flux of $\vec{D}$ out of a closed surface S = Total free charge enclosed within S.

That is the surface integral of the normal component of electric displacement vector $\vec{D}$ over any closed surface equals the charge enclosed, *where Q is the free charge and the polarisation charge is not included. Equation (11) can be* written in divergence form as follows. Since according to divergence theorem,

$$\int_S \vec{D}.dS = \int_V \text{div}\,\vec{D}\,dv$$

and $$Q = \int_V \rho_f \, dv$$

hence, $$\vec{\nabla} \cdot \vec{D} = \rho f \qquad ...(12)$$

where ρf is the volume density of free charge Q. For physical significance see the Gauss's law, equation (12) is still really Gauss's law, now modified in such a way that the effects of polarisation charge are automatically included.

Verification of relation $\vec{D} = \hat{I}_0 \vec{E} + \vec{P}$

The variation of electric displacement $\vec{D}$ electric field intensity $\vec{E}$, polarisation $\vec{P}$ along the axis between the plates of an air capacitor. The variation after a dielectric (paraffin) slab has been introduced is illustrated. For a given applied potential *i.e.,* battery is disconnected when the paraffin is introduced. *In air capacitor reduces,*

$$\vec{D} = \epsilon_0 \vec{E} \qquad ...(13)$$

because there is no polarisation, hence,

$$\vec{P} = 0, \text{ and } \vec{E} = \frac{\vec{D}}{\epsilon_0} \qquad ...(14)$$

In paraffin capacitor the relation,

$$\vec{D} = \epsilon_0 \vec{E} + \vec{P}$$

The relation between $\vec{D}$ and $\vec{E}$ for this case.

$$\vec{D} = K \epsilon_0 \vec{E}$$

$$\vec{E} = \frac{\vec{D}}{\epsilon_0 K} \qquad ...(15)$$

and the polarisation $\vec{P}$ by,

$$\vec{P} = \epsilon_0 (K - 1) \vec{E}$$

or $$\vec{P} = \epsilon_0 K\left(1-\frac{1}{K}\right)\vec{E}$$

$$= \vec{D}\left(1-\frac{1}{K}\right) \quad ...(16)$$

K for paraffin is 2.1, hence the capacitance with paraffin increases by 2.1 times. That is with paraffin dielectric the capacitor plates can store 2.1 times charge for a given potential. *Since the battery was disconnected before the paraffin was introduced, the charge on the plates is same as with air.* Therefore, $\vec{D}$ is the same as before. However, now electric intensity,

$$\vec{E} = \frac{\vec{D}}{\epsilon_0 K} = 0.475\frac{\vec{D}}{\epsilon_0}, \text{with air it is } \frac{\vec{D}}{\epsilon_0} \quad ...(17)$$

the polarisation $\vec{P} = \vec{D}\left(1-\frac{1}{K}\right) = 0.525\ \vec{D}$, with air $\vec{P} = 0$ ...(18)

However, the polarisation $\vec{P}$ has such value that

$\vec{D}$ with dielectric $= \vec{P} + \epsilon_0 \vec{E} = 0.525\ \vec{D} + 0.475\ \vec{D} = \vec{D}$ ...(19)

$\vec{D}$ with air $= \epsilon_0 \vec{E} = \epsilon_0 \frac{\vec{D}}{\epsilon_0} = \vec{D}$

Hence is verified.

Mechanism of Polarisation

Since α is defined in terms of dipole moment, its magnitude will clearly be a measure of the extent to which electric dipoles are formed by ions, atoms and molecules. The *electric dipoles are formed through a variety of mechanisms,* each one contributing to the value of α, hence α is regarded as total polarisability to be the sum of individual polarisabilities each arising from a particular mechanism. In general, the molecules or atoms (ions) may be affected in the following ways :

ELECTRONIC POLARISABILITY

Centres of gravity of the electrons and protons may change resulting in induced dipole moment *i.e.,*

$$P_e = \alpha_e E_{int}$$

P_e has all the characteristics of an assembly of dipoles produced by displacement of electrons which have natural frequency equal to, or higher than, those of visible light. So that α_e is called the *optical or electronic polarisability.*

MOLECULAR POLARISABILITY

Molecules having permanent dipole moment *i.e.,* polar molecules are affected in two possible ways:

(a) Atomic Polarisability

Firstly, the field may cause the atoms to be displaced altering the distance between them and hence changing the dipole moment of the molecule. Such polarisability is called atomic represented by αa.

(b) *Orientational or Dipolar Polarisability*

Secondly, the molecule as a whole may rotate about its axis of symmetry, so that the dipoles align with the field. This is referred as orinentational polarisation or dipolar polarization and polarisability is called orientational or dipolar, denoted by α_d.

INTERFACIAL POLARISABILITY

Such type of mechanism is due to large number of defects in the structure of crystals, such as lattice vacancies, impurity centre, dislocation, etc. Polarisability is denoted by α_i, however, this is generally neglected. Thus we can write the total polarisability α as follows :

$$\alpha = \alpha_e + \alpha_a + \alpha_d + \alpha_i \qquad ...(1)$$

Lorentz Local Field

In *i.e.,* $P = N \alpha E_{int}$, it should be noted that this internal intensity of the field, where molecules exist surrounded on all sides by other polarised molecules of matter is, not equal to the external intensity of the field at a given point of a dielectric which is calculated by relation $E = V/d$. The internal intensity E_{int} on a molecule is a geometrical sum of the external intensity E and the component of the intensity caused by the action of the other polarised molecules of the matter on the molecule in question. The following method suggested by Lorentz can be used to estimate the relation between E and E_{int}.

Let us describe a sphere in a dielectric, containing the molecule considered above in its centre. The molecules present inside dielectric outside of the sphere will build up an intensity component E_1 in the centre of the sphere and those inside the sphere an intensity component E_2. Thus internal intensity will be equal to :

$$\vec{E}_{int} = \vec{E} + \vec{E}_1 + \vec{E}_2 \qquad ...(2)$$

To calculate E_1, let us imagine that the dielectric is removed from the sphere. For the actual pattern of the electric field not to be distorted, a surface electric charge should be placed on the sphere.

At each point of the sphere the surface density of the charge is given by, *i.e.*, $\sigma = \vec{P}.\hat{n}dS$ that is the definition for $\vec{P}$ so that,

$$\sigma = P \cos \theta \qquad ...(3)$$

where θ is the angle between the direction of $\vec{P}$ *i.e.*, the direction of $\vec{E}$ and the radius of the sphere. The charge dq on the small element of the sphere surface dS will be given by,

$dq = \sigma$ dS, substituting from Eq. (3), we get,

$$dq = P \text{ Cos } \theta dS \qquad ...(4)$$

The field created by this charge dq inside the sphere,

$$dE_1 = \frac{1}{4\pi \in_o} \frac{dq}{r^2}$$

Putting dq from equation (4),

$$dE_1 = \frac{P}{4\pi \in_o r^2} \cos \theta \, dS \qquad ...(5)$$

The field dE_1 has the component one is parallel to the direction of E.

$$dE_1 \text{ Cos } \theta = \frac{P}{4\pi \in_o r^2} \text{Cos}^2 \theta \, dS \qquad ...(6)$$

And the other perpendicular to the direction of E *i.e.*,

$$dE_1 \text{ Sin } \theta = \frac{P}{4\pi \in_o r^2} \cos \theta \text{ Sin}\theta \, dS \qquad ...(7)$$

The total intensity E_1 of the entire charge on the sphere is determined by integrating dE_1 over the whole surface of the sphere. But the components of the upper and lower halves of the sphere perpendicular to E will cancel. So there will be only parallel components. Therefore, the component of E_1 at O due to the right half sphere,

$$= dE_1 \text{ Cos } \theta = \int_0^{\pi/2} \frac{P}{4\pi \epsilon_o} \frac{\text{Cos}^2 \theta}{r^2} dS$$

dS is the area or right ABCD *i.e.,*

$$dS = 2\pi r^2 \text{ Sin } \theta \, d\theta$$

$$\text{Therefore, } dE_1 \text{ Cos } \theta = \int_0^{\pi/2} \frac{P}{4\pi \epsilon_o} \frac{\text{Cos}^2 \theta}{r^2} 2\pi r^2 \text{ Sin } \theta \, d\theta$$

$$= \frac{P}{2 \epsilon_o} \int_0^{\pi/2} \text{Cos}^2 \theta \text{ Sin } \theta \, d\theta$$

$$= \frac{P}{2 \epsilon_o} \left[-\frac{\text{Cos}^3\theta}{3} \right]_0^{\pi/2}$$

$$= \frac{P}{6 \epsilon_o} \quad \text{...(8)}$$

As the charges on the surfaces of the two halves of the sphere are equal in magnitude but opposite in sign an equal force acting in the same direction will be exerted on unit positive charge at O. So, the component of E_1 due to left half sphere is also given by,

$$= \frac{P}{6 \epsilon_o}$$

So, the total intensity E_1 is sum of the two i.e,

$$\vec{E} = \frac{\vec{P}}{3 \epsilon_o} \quad \text{...(9)}$$

The direction of this intensity coincides with the direction E, *the value of E_l given is called Lorentz field.*

Clausius Mosotti Equation

For the simplest case when there is no prevailing effect of any adjacent charges on the molecules in question *i.e.,* $\vec{E}_2$ components of taken zero,

$$\vec{E}_{int} = \vec{E} + \frac{\vec{P}}{3\epsilon_o} \qquad ...(10)$$

We have also a relation between $\vec{P}$ and $\vec{E}$, *i.e.,*

$$P = \epsilon_o (K - 1) \vec{E}$$

or
$$\frac{\vec{P}}{\epsilon_o} = (K-1) \vec{E}$$

inserting $\frac{\vec{P}}{\epsilon_o}$ in equation (10), we get,

$$\vec{E}_{int} = \vec{E} \left(1 + \frac{(K-1)}{3}\right)$$

$$\vec{E}_{int} = \vec{E} \left[\frac{K+2}{3}\right] \qquad ...(11)$$

The $\vec{E}_{int}$ given in equation (11) is called Mosotti field. Since the expression for $\vec{E}_{int}$ contains K, hence it depends on dielectric material in case of vacuum capacitor with K = 1, that the ratio $\frac{K+2}{3} = 1$ and $\vec{E}_{int} = \vec{E}$.

The value of polarisation P as an electric moment of a unit volume of dielectric *i.e.,*

$$\vec{P} = N \alpha \vec{E}_{int}$$

Putting the value $\vec{E}_{int}$ from equation (11), in the above equation, we have,

$$\vec{P} = N\alpha \frac{(K+2)}{3} \vec{E}$$

and equate this expression to P given below:

$$\vec{P} = \epsilon_o (K - 1) \vec{E},$$

i.e.,
$$(K - 1) \epsilon_o E = N \alpha E \frac{K+2}{3}$$

or
$$\frac{K-1}{K+2} = \frac{N\alpha}{3\epsilon_o} \qquad ...(13)$$

where $\alpha = \alpha_e + \alpha_a$. Equation (13) is *remarkable for the fact that it shows the connection between the parameters K of a dielectric and its parameters α and N.* The equation (13) *is known as Clausius Mosotti equation* As $K = \epsilon / \epsilon_o$ the Clausius Mosotti equation in another simple form is,

$$\frac{\epsilon - \epsilon_o}{\epsilon + 2\epsilon_o} = \frac{N\alpha}{3\epsilon_o} \qquad ...(14)$$

The magnitude $\dfrac{\epsilon - \epsilon_o}{\epsilon + 2\epsilon_o}$ or $\dfrac{K-1}{K+2}$ is known as *specific polarisation of a dielectric.* This is dimensionless because K is dimensionless. It is easy to see that the R.H.S. of equation (13) is also dimensionless because N is measured in meter^{-3}, α in farad meter^2 and ϵ_o in farad/meter, hence $N\alpha/3\epsilon_o$ is dimensionless.

Now using relation $K = 1 + \chi_{er}$. *The Clausius Mosotti equation reduces in terms of the susceptibility of the dielectric.*

$$\frac{\chi_{er}}{\chi_{er} + 3} = \frac{N\alpha}{3\epsilon_o} \qquad ...(15)$$

A version of Clausius Mosotti equation convenient for practical calculation is obtained by multiplying equation (13) on both sides by the ratio of the molecular mass M of the substance to its density D.

$$\frac{K-1}{K+2}\frac{M}{D} = \frac{N\alpha}{3\epsilon_o}\frac{M}{D}$$

Since $\dfrac{NM}{D}$ = Avogadro number N_a

$$\frac{K-1}{K+2}\frac{M}{D} = \frac{1}{3\epsilon_o}N_a\alpha \qquad ...(16)$$

but $\dfrac{K-1}{K+2}$ = specific polarisation (P_{sp})

Hence, equation (16) reduces,

$$\frac{K-1}{K+2}\frac{M}{D} = P_{sp}\frac{M}{D} \qquad ...(17)$$

The magnitude given in equation (17) is called the *molar polarisation represented by π,*

$$\pi = \frac{K-1}{K+2}\frac{M}{D} = P_{sp}\frac{M}{D} = \frac{1}{3\epsilon_0}N_a\alpha \qquad ...(18)$$

The dimension of molar polarisation is equal to that of ratio of volume to mass, *i.e.,* the dimension inverse to that of density. For neutral dielectrics the value of K η^2 where η is the refractive index of the medium then we get after substituting in equ. (17) and (18),

$$\frac{\eta^2-1}{\eta^2+2} = \frac{N\alpha}{3\epsilon_0} \qquad ...(19)$$

and
$$\frac{\eta^2-1}{\eta^2+2}\frac{M}{D} = \frac{N_a\alpha}{3\epsilon_0} \qquad ...(20)$$

Equations (19) and (20) are referred *as Lorentz-equation.*

LIMITATIONS OF THE CLAUSIUS MOSOTTI EQUATION

This equation is derived by taking $E_2 = 0$ for details of E_2 because of the following assumptions:

1. Since polarisation is considered as proportional to the field, it means the polarisation of the molecules by elastic displacement only.
2. Absence of short-range interaction.
3. Isotropy of the polarisability of the molecules.

All these conditions are satisfied with neutral molecules having no constant dipoles *i.e., non polar* (molecules of symmetrical structure). Thus the equation is applicable to *neutral liquids and specially to gases in which the molecules are far apart.*

If a graph is plotted between $\frac{K-1}{K+2}\frac{1}{D}$ on one axis and $\eta = \sqrt{K}$ on the other axis with different values of D, for D = 1, gives the function $\frac{K-1}{K+2}$ (K) approximating unity when K infinitely increases. *i.e.,*

$$\frac{K-1}{K+2} \cong 1 \qquad ...(1)$$

Since K in gases is close to unity, we can write,

$$\frac{K-1}{K+2} \cong \frac{K-1}{3}$$

Equation (4.73) reduces,

$$\frac{K-1}{K+2} \cong \frac{K-1}{3} = \frac{N\alpha}{3\in_0}$$

or $$K = 1 + \frac{N\alpha}{3\in_0} \qquad ...(2)$$

Debye Equation

When a *polar molecule* is placed in the electric field, it is affected in two ways. *Firstly,* it displaces the centre of gravity of protons and electron, so that an extra dipole moment is induced, giving the *electronic* (atomic) polarisability. For which the *Clausins Mosotti equation is valid. Secondly, the dipole tend to orient* itself so that its potential energy is minimum *i.e.,* it will *tend to align its moment* against the field.

The process is called *orientational polarisations. Debye transformed the Clausius-Mosotti equation for polar molecules* by replacing the electronic and ionic (or atomic) polarisability α of the molecule which is taken in Clausius-Mosotti equation with the sum of deformational polarisability α and the polarisability due to dipole polarisation α_d.

The Clausius-Mosotti equation has no indication of a *temperature variation of permittivity.* However, *for polar dielectrics the polarisability is inversely proportional to temperature due to thermal agitation.* So that orientational polarisation mechanism is temperature dependent. In fact the orientative action of the field is hindered by the thermal agitation. In consequence, a statistical equilibrium is set up in which a slight excess of molecules have their permanent dipoles antiparallel to the field, *i.e.,* the molecules are aligned.

Let us calculate the average component of dipole moment per molecule at temperature T due to alignment. If molecule of permanent dipole moment $\vec{P}_o$ has its axis inclined at an angle θ with the field, then its potential energy is,

$$U = -\vec{P}_o \cdot \vec{E}, \qquad ...(3)$$

According to Maxwell Boltzmann law, the number of molecules distributed with axis of their dipoles pointing in all direction within a solid angle d Ω at θ is,

$$dn = A \exp\left[-\frac{U}{K_b T}\right] d\Omega \qquad ...(4)$$

where K_b is Boltzmann's constant, and A is constant depending on the number of molecules considered.

Integrating this overall possible directions, total number of molecules after putting the value of U from equation (3),

$$n = \int_{\theta=0}^{\theta=\pi} A \exp\left[\frac{P_o E \cos\theta}{K_b T}\right] d\Omega \qquad ...(5)$$

The effective dipole moment *i.e.,* the dipole moment of a dipole along the field direction = P_o Cos θ. Therefore, the dipole moment of dn molecules in the field direction,

$$= dn\, P_o \cos\theta$$

$$= A \exp\left[\frac{P_o E \cos\theta}{K_b T}\right] P_o \cos\theta\, d\Omega,$$

Integrating this within limit $\theta = 0$ to $\theta = \pi$, we get the total dipole moment of all molecules,

$$\sum P_o = \int_{\theta=0}^{\theta=\pi} A e^{x\cos\theta} P_o \cos\theta\, d\Omega$$

where, $$x = \frac{P_o E}{K_b T}$$

The *average dipole moment* ($\overline{P}_o$) along the field,

$$\overline{P}_o = \frac{\sum P_o}{n} = \frac{\int_0^{\pi} A e^{x\cos\theta} P_o \cos\theta\, d\Omega}{\int_0^{\pi} A e^{x\cos\theta}\, d\Omega} \qquad ...(6)$$

Now the value of $d\Omega$ between θ and $\theta + d\theta$ is calculated.

$$d\Omega = \frac{\text{area of ring between } \theta \text{ and } \theta + d\theta}{r^2}$$

$$= \frac{2\pi r \sin\theta \, r d\theta}{r^2}$$

$$= 2\pi \sin\theta \, d\theta \qquad \text{...(7)}$$

to evaluate the integral of equation (7) we put,

Cos $\theta = \alpha$. as θ varies from 0 to π, a will vary + 1 to – 1

Hence. Sin $\theta \, d\theta = - d\alpha$. therefore,

Substituting in equation (7), this value we get,

$$\overline{P}_o = \frac{\int_{-1}^{+1} e^{xa} P_o a da}{\int_{-1}^{+1} e^{xa} da}$$

$$\frac{\overline{P}_o}{P_o} = \frac{\left[\frac{a}{x} e^{xa} - \frac{1}{x^2} e^{xa}\right]_{-1}^{+1}}{\left[\frac{1}{x^2} e^{xa}\right]_{-1}^{+1}}$$

$$= \frac{e^x + e^{-x}}{e^x - e^{-x}} - \frac{1}{x}$$

$$= \coth x - \frac{1}{x}$$

therefore, $$\frac{\overline{P}_o}{\overline{P}_o} = L(x) \qquad \text{...(8)}$$

where L (x) = coth x –1/x is called *Langevin function,* because it was derived by Langevin. The relation (8) is called *Langevin formula.* The variation of L (x) with x.

The curve is linear for small values of x, *i.e.,* for very small value of E and large values of t, where for large value of x *i.e.,* $\frac{P_o E}{K_b T}$, it approaches unity. Therefore, for *large value of E saturation is obtained i.e., practically all the molecules are aligned with field direction and further increase in E has on effect on average dipole moment.* However, the linear region corresponding to small values of x is important at ordinary temperature. The dipole moment P_o of most polar materials is such that x < < 1 for a full range of field strength, even fro those approaching the dielectric strength of the material, so long as the temperature is above about 250° K. Thus a polar dielectric is, in general linear. For small values of x, L (x) can be expanded as,

$$L(x) = \frac{x}{3} - \frac{x^3}{4 \times 5} + \ldots$$

Practically x is very small, hence,

$$L(x) = \frac{x}{3}$$

Hence, equation (8) reduces,

$$\frac{\overline{P}_o}{\overline{P}_o} = \frac{x}{3}$$

$$= \frac{P_o E}{3K_b T}$$

or $$\overline{P}_o = \frac{P_o^2 E}{3K_b T} \qquad ...(9)$$

E is the polarising field *i.e.*, the field acting on the dipoles. Comparison of this with *i.e.*, $P = \alpha E_{int}$ which is for linear dielectric, E_{int} here is E, we get polarisability α_d (*i.e.*, molecular dipole moment per unit polarising field) is,

$$\alpha_d = \frac{P_o^2}{3K_b T} \qquad ...(10)$$

This represent the orientational polarisability. Therefore, the total polarisability

$$\alpha_{eff} = \alpha + \alpha_d = \alpha + = \frac{P_o^2}{3K_b T} \qquad ...(11)$$

where α is deformational polarisability as discussed in the beginning ($\alpha = \alpha_e + \alpha_a$). Hence,

$$\alpha_{eff} = \alpha_e + \alpha_a + \frac{P_o^2}{3K_bT}$$

Is known as the *Langevin-Debye equation.* Therefore, in the case of polar molecule replacing α by α_{eff} as below:

$$P_{sp} = \frac{\epsilon - \epsilon_o}{\epsilon + 2\epsilon_o} = \frac{K-1}{K+2} = \frac{N}{3\,\epsilon_o}\left(\alpha + \frac{P_o^2}{3K_bT}\right) \quad ...(12)$$

and

$$\pi = \frac{\epsilon - \epsilon_o}{\epsilon + 2\epsilon_o}\frac{M}{D} = \frac{K-1}{K+2}\frac{M}{D}$$

$$= \frac{N_\alpha}{3\,\epsilon_o}\left(\alpha + \frac{P_o^2}{3K_bT}\right) \quad ...(13)$$

Temperature Dependence

Equation (11) shows that α_{eff} depends on temp T. Therefore *for polar molecule the polarisability is inversely proportional to the temperature so we can say that K and permittivity also depends on temperature in the same manner as α.* (Eqn. (13)).

Since $\frac{P_o^2}{3K_bT} >> \alpha$, $\alpha v \delta$ for *this reason the permittivity of polar dielectric may be much higher than that of non-polar.* The temperature dependence of dielectric constant K. The intercept of the straight line thus obtained on the y-axis is $\frac{N_\alpha}{3\,\epsilon_o}$ and it slope is $\frac{N_\alpha P_o^2}{9\,\epsilon_o\,K_b}$

FERROELECTRIC AND PARAELECTRIC DIELECTRIC

The earlier classification of dielectrics into *polar and non-polar is not sufficient.* They *are classified according to their behaviour in electric field.* Take the equation $\vec{P} = \epsilon_o (k-1)\,\vec{E}, \vec{E}$ is the resultant of polarising (external field) E_o and field due to polarisation charges E', *i.e.*, $\vec{E}_o + \vec{E}'$ and $\vec{E}$ is related to E_o by relation,

$$\vec{E} = \vec{E}_0 /K$$

In most cases the polarisation is proportional to $\vec{E}$.

When polarising field $\vec{E}_0$ is removed then $\vec{E}$ vanishes and in turn $\vec{P}$ also vanishes. But under certain conditions given by equation (4.71),

$$E_{int} = E + \frac{P}{3 \epsilon_0}$$

Now when external field is removed, E is zero, then

$E_{int} = \frac{P}{3 \epsilon_0}$ [P_o is polarisation without external field called permanent polarisation]

Ferroelectric

So we can say that *certain materials, retain a permanent polarisation even though the polarising field is removed, such materials are called ferro-electric by analogy with ferromagnetism.*

In other words if a polarisation P_0 exists, it will create an electric field at the molecule which tends to polarise the molecules. Such polarisation is called *spontaneous polarisation.* Such materials are able to develop large polarisation when placed in an electric field K >> 1 and $\chi er > 1$.

Examples : Rochelle salt, $BaTiO_3$ (Barium Titenate), $SrTiO_3$ (Stroncium titenate), $KNbO_3$ (Potensium niobate).

Paraelectric

Materials with *small positive susceptibility* due to polarisation in an electric field are termed as para-electric. K and χ_{er} are positive but having small value than ferroelectric. In fact all dielectrics are para with the exception of vacuum, which has precisely zero susceptibility χ_{er} = K – 1 , for vacuum K = 1.

Dia-electric do not Exist

There is no material which has *negative susceptibility.* There is no *analogy with diamagnetism.* In other words, the polarisation in the direction opposite to the polarising field is not possible.

Hysteresis Loop for Ferroelectric

Complete curve of *polarisation us, electric field.* As it is the property of ferro-electric to retain polarisation when polarising field is removed, a curve such as in Fig. 4.23 is called *hysteresis loop. Hysteresis means to lag behind,* and it is clear that $\vec{p}$ *lags the electric field vector* $\vec{E}$ when $\vec{E} = 0$, the retained polarisation is ob, it is *spontaneous polarisation (analogous to residual magnetism in magnetisation).* And point C is the electric field which is needed to *completely reverse the polarisation, it is analogous to coercive force in magnetisation.*

CURIE TEMPERATURE

However, the *spontaneous polarisation exists only below the certain critical temperature called Curie temperature. Above the* Curie temperature, the *spontaneous polarisation disappears and* substance becomes *para electric.* Below the critical temperature, the variation of susceptibility with temperature is governed by Curie law and above critical temperature by curie Weiss law.

So, we can say that polarisation study of a dielectric is more or less analogous to magnetisation: The behaviour of materials in electric and magnetic field are identical.

Energy and Force Relations for Electrostatics Field with Dielectric Present

The basic energy and pressure laws are modified with dielectrics. When a dielectric is placed in an external field, it is polarised. Let the induced charge + q appears on one surface and – q on the opposite surface, + q and – q are separated by distance d is the case of parallel plate capacitor with dielectric *i.e.,* the thickness of the dielectric slab, Dipole moment $\vec{p} = qd$. Suppose that while the field $\vec{E}$ is present the charges are further displaced by a distance Δd. The change in dipole moment is given by,

$$dp = q\,(d + \Delta d) - qd$$

or

$$dp = q\,\Delta\, d.$$

The force on the polarisation charge is qE. So the work done for displacement Δ d is given by,

$$dW = q\,E\,\Delta\, d$$

Δd is the displacement in the direction of $\vec{E}$ inserting the value of dP given above, we get,

$$dW = \vec{E}.d\vec{p} \text{ as } \vec{E} \text{ and } \vec{P} \text{ point in the same direction.}$$

This work done is stored in the form o of energy in the dielectric. The supplier of the work is the source of the field, the battery that maintained constant potential difference, between the plates. This stored energy in the dielectric increases the internal energy of the molecules.

But since $\vec{P} = \epsilon_0 (K - 1) \vec{E}$, for linear dielectric, where $\vec{P}$ is dipole moment per unit volume. Therefore,

$$d\vec{P} = \hat{I}_0 (K - 1)\, d\vec{E}$$

where $d\vec{P}$ is the change in dipole moment per unit volume. So work done is also given by,

$$\vec{E}.d\vec{P} = \vec{E}.\epsilon_0 (K - 1)\, d\vec{E}$$

Since, $\vec{E}.d\vec{E} = EdE = 1/2\, d\,(E^2)$, therefore,

$$\vec{E}.d\vec{P} = \frac{\epsilon_0}{2} (K - 1)\, d\,(E^2)$$

or

$$\int_0^P \vec{E}.d\vec{P} = \frac{\epsilon_0}{2} (K - 1)\, E^2$$

This equation gives the energy change due to polarisation P. Therefore the *total energy density equal to energy density in free space plus the energy density in dielectric.* Therefore, the total energy density,

$$= \frac{1}{2} \epsilon_0 E^2 + \frac{\epsilon_0}{2} (k-1) E^2$$

$$= \frac{1}{2} \epsilon_0 KE^2 \qquad ...(1)$$

That is *energy increases by K times than that of without dielectric.* Energy per unit volume represented by U is given by,

$$U = \frac{\epsilon_0}{2} KE^2 \text{ joules/meter3}$$

Since $\vec{D} = K \epsilon_0 \vec{E}$ and $\vec{D}$ is in the direction of $\vec{E}$, therefore we can write,

$$U = \frac{1}{2}\ \vec{E}.\vec{D} \text{ joules/meter}^3 \qquad ...(2)$$

For any volume,

$$U = \frac{1}{2}\ \epsilon_o \int_v KE^2 dv \qquad ...(3)$$

or $$U = \frac{1}{2} \int_v \vec{E}.\vec{D}\ dv \qquad ...(4)$$

If *dielectric is isotropic* then,

$$U = \frac{1}{2}\ \epsilon_o K \int_v E^2\ dv \text{ joules} \qquad ...(5)$$

Though we have *derived it for parallel plate capacitor but it is applicable in all cases.*

Force

An expression for force can be obtained by *principle of virtual work.* If the work is done by an external agent against the force of attraction between the plates in moving them further apart through a distance Δd,

$$W = \vec{F}.\Delta\vec{d}$$

or $$W = -F\ \Delta\ d$$

The change in volume due to this additional separation. Δd = A Δd, where A is the area of the plate, change in energy due to this change in volume equals $\frac{1}{2}\ \epsilon_o\ KE^2\ A\ \Delta d$

work done = change in energy

$$\frac{1}{2}\ \epsilon_o\ KE^2\ A\ \Delta d = \vec{F}.\Delta\vec{d}$$

$$\frac{1}{2}(\vec{E}.\vec{D})A\ \Delta d = \vec{F}.\Delta\vec{d}$$

Since $\vec{D} = \epsilon_o\ K\ \vec{E}$ and in the direction of $\vec{E}$

or $$F = \frac{1}{2}\vec{E}.\vec{D}\ A \text{ Newtons}$$

(Since , $\vec{E}$ = Newtons/coul, $\vec{D}$ = Couls/meter^2 and A = meter^2

$$\text{Pressure} = = \frac{\vec{F}}{A} = \frac{1}{2}\vec{E}.\vec{D} \text{ Newtons/meter}^2 \qquad ...(6)$$

This is force of attraction between the plates tending to pull them together. Though the pressure exists at any point in an electric field.

PRESSURE AT A CONDUCTOR (PLATES)—DIELECTRIC BOUNDARY

The pressure is in the direction to draw the conductor into the dielectric. The result follows because the two plates attract each other. The pressure of the conductor on the dielectric is given by equation (6) which may be written in several forms,

$$P = \frac{1}{2}\ \vec{E}.\vec{D}\ \frac{1}{2}\ \frac{D^2}{\epsilon} \text{ since } \vec{D} = \epsilon_0\ K\ \vec{E} \text{ and } K = \frac{\epsilon_0}{\epsilon}$$

$$= \frac{1}{2} \in E^2$$

$$= \frac{1}{2}\ \frac{\sigma^2}{\epsilon} \qquad ..(1)$$

Since σ is surface density of charge and given by $\sigma = K\ \epsilon_0\ \ E = D$

Alternate Method

As we discussed in Chapter 5, that the potential energy due to a charge distribution is given by,

$$U = \frac{1}{2}\int_v \rho\ V\ dv$$

where v is the volume.

Now according,

$$\vec{\nabla}.\vec{D} = \rho$$

$$\text{Therefore,} \quad U = \frac{1}{2}\int_v V\ (\nabla.\vec{D})\ dv \qquad ...(2)$$

$$\text{but,} \quad \nabla.\ (V.\vec{D}) = V\ (\nabla.\vec{D}) + \vec{D}.\nabla V,$$

Substituting in above equation (2), we get,

$$U = \frac{1}{2}\int_V \nabla.(V.\vec{D})\, dv - \frac{1}{2}\int_V \vec{D} \text{ grad } V\, dv,$$

(Since (∇ V) = grade V). But according to divergence theorem,

$$\int_V \nabla.(V.\vec{D})\, dv = \int_S V\vec{D}.d\vec{S}$$

therefore, $$U = \frac{1}{2}\int_S V\vec{D}.d\vec{S} - \frac{1}{2}\int_V \vec{D}.\text{gard } V\, dv$$

As we know that $V \alpha \frac{1}{r}$

$$D \alpha \frac{1}{r^2}$$

$$dS \alpha r^2$$

So, $$\vec{V}.D\, dS \alpha \frac{1}{r}$$

Hence $\int_S VD\, dS$ *will vanish if the ;surface be expanded infinitely, therefore,*

$$U = \frac{1}{2}\int_V \vec{D}.\vec{E}\, dv, \quad (\text{as } E = -\text{ grad } V) \qquad ...(3)$$

SOLVED EXAMPLES

Example 1:

The plates of parallel plate capacitor are separated by 1.00 mm. What must be the plate area if the capacitance to be 1.0 Farad assuming that air is filled between the plates.

Solution:

Using equation (1), we get,

$$A = \frac{dC}{\epsilon_o} = \frac{1.0\times10^{-3}\text{ metre } (1.00 \text{ Farad})}{8.9\times10^{-12}\text{ coul}^2/\text{newton. metre}^2}$$

as, $$\text{Farad} = \frac{\text{Coul}}{\text{Volt}} = \frac{\text{Coul}^2}{\text{newton.metre}}$$

Hence, $A = 1.1 \times 10^8 \text{ metre}^2$

This is approximately the area of a square sheet of more than 9 km of side. Indeed farad is very large unit. Therefore, we use smaller unit as micro farad or micro-micro farad.

Example 2:

A parallel plate capacitor is filled with two dielectrics of same dimensions but different dielectric constant K_1 *and* K_2 *respectively. calculate its capacitance.*

Solution:

From the arrangement is simply two capacitors in parallel, and each capacitor will have the area A/2.

Total capacitance $C = C_1 + C_2$

$$= \frac{\epsilon_0 A/2k_1}{d} + \frac{\epsilon_0 A/2k_2}{d}$$

$$= \frac{\epsilon_0 A}{d}\left(\frac{k_1 + k_2}{2}\right)$$

The arrangement of can be regarded as two capacitors in series. Each capacitor will be of area A and separation d/2. Therefore the total capacitance is given by the relation,

$$\frac{1}{C} = \frac{1}{C_1} + \frac{1}{C_2} = \frac{d}{2}\left(\frac{1}{\epsilon_0 AK_1} + \frac{1}{\epsilon_0 AK_2}\right)$$

or $$C = \frac{2\epsilon_0 A}{d}\left(\frac{K_1K_2}{K_1 + K_2}\right)$$

Example 3:

The separation between the plates of parallel plate capacitor is d, one plate is at zero potential. The space between the plates is filled with a material whose dielectric constant varies uniformly from one plate to the other. If K_1 *and* K_2 *are the values of the dielectric constant at the two plates, show that the capacity per unit area is*

$$C = \frac{\epsilon_o}{d} \frac{K_2 - K_1}{\log_e \frac{K_2}{K_1}}$$

Solution:

The electric intensity E is given by,

$$E = \frac{\sigma}{\epsilon_o K}, \text{ since } E = \frac{E_o}{K}$$

As K is not constant, but varies from K_1 to K_2 increase in dielectric constant per unit length $= \frac{K_2 - K_1}{d} = \alpha$. The dielectric constant at point at a distance x from the first plate is $K = K_1 + \alpha x$.

Hence the electric intensity at this point $= \frac{\sigma}{\epsilon_o (K_1 + \alpha x)}$

Potential difference between the plates,

$$V = -\int_{\alpha}^{0} \vec{E}.d\vec{x} = \int_{0}^{d} E dx$$

$$= \frac{\sigma}{\alpha \epsilon_0} \log \frac{K_1 + \alpha d}{K_1}, \text{ substituting } \alpha$$

$$= \frac{\sigma d}{\epsilon_0 (K_2 - K_1)} \log \frac{K_1 + \frac{(K_2 - K_1) d}{d}}{K_1}$$

$$= \frac{\sigma d}{\epsilon_0 (K_2 - K_1)} \log_e \frac{K_2}{K_1}$$

Hence, $C = \frac{\text{total charge}}{V} = \frac{\sigma A}{V}$

$$= \frac{A \epsilon_o (K_2 - K_1)}{d \log_e K_2 / K_1}$$

Example 4:

A parallel plate capacitor consists of two square metal plates 5.0cm of side and separated by 1 cm. A sulphur slab 6 mm thick is placed on the lower plate, calculate the capacitance of the capacitor, dielectric constant of sulphur is 4.

Solution:

When sulphur slab of 6 mm is placed on the lower plate, it leaves an air gap of 4 mm thickness. The capacitor can be regarded as two capacitors in series.

An air capacitors of 4 mm plate spacing and a sulphur field capacitor of 6 mm plate space. From capacitance of air capacitor,

$$C_a = \frac{\epsilon_0 A}{d} \text{ farad when A and d are in metres}$$

$$= 8.85 \times 10^{-12} \frac{A}{d} \text{ farad}$$

$$= 8.85 \frac{A}{d} \quad \mu\mu F,$$

since $A = 0.5 \times 0.5$
and $d = 4 \text{ mm} = 0.004$ metre

hence, $= \dfrac{8.85 \times 0.5^2}{0.004}$

$$= 553 \ \mu\mu F$$

The capacitance of the sulphur filled capacitor,

$$C_s = \frac{K \epsilon_0 A}{d'}$$

$$= \frac{8.85 \times 4 \times 0.5^2}{0.006} \mu\mu F$$

$$= 1{,}475 \ \mu\mu F$$

Then total capacitance,

$$\frac{1}{C} = \frac{1}{C_a} + \frac{1}{C_s}$$

$$C = \frac{C_a C_s}{C_a + C_s} = 402 \ \mu\mu F$$

Example 5:

A concentric spherical capacitor is formed is formed by two concentric spheres of radii a and b. If the medium between the spherical shells has the dielectric constant K_1 from a or r and K_2 from r to b show that

$$C = \frac{4\pi \in_o}{\left[\frac{1}{k_1 a} - \frac{1}{k_2 b} + \frac{1}{r}\left(\frac{1}{k_2} - \frac{1}{k_1}\right)\right]}$$

Solution:

The potential difference between the inner and outer spheres is given by,

$$V_a - V_b = -\int_b^a E dr = +\int_a^b \frac{Q}{4\pi \in_o Kr^2} dr$$

$$= \frac{Q}{4\pi \in_o} \int_a^b \frac{dr}{Kr^2}$$

AS $K = K_1$ from a to r and $K = K_2$ from r to b,

$$V_a - V_b = \frac{Q}{4\pi \in_o}\left[\int_a^r \frac{dr}{K_1 r^2} + \int_r^b \frac{dr}{K_2 r^2}\right]$$

$$= \frac{Q}{r\pi \in_o}\left[\left|-\frac{1}{K_1 r}\right|_a^r + \left|-\frac{1}{K_2 r}\right|_r^b\right]$$

$$= \frac{Q}{4\pi \in_o}\left[\frac{1}{r}\left(\frac{1}{K_2} - \frac{1}{K_1}\right) + \left(\frac{1}{K_1 a} - \frac{1}{K_2 b}\right)\right]$$

Therefore the capacitance

$$C = \frac{Q}{V_a - V_b}$$

$$= \frac{4\pi \in_o}{\frac{1}{r}\left(\frac{1}{K_2} - \frac{1}{K_1}\right) + \left(\frac{1}{K_1 a} - \frac{1}{K_2 b}\right)}$$

Example 6:

For a parallel plate capacitor following data are given—area of the each plate = 100 cm^2*; separation of plates = 1.00 cm A potential difference of 100 volt is applied with no dielectric present. A dielectric slab of dielectric constant 7.00 and of thickness 0.50 cm is introduced. Calculate:*

(1) the free charge

(2) electric field strength in the gap,

(3) electric field strength in the dielectric,

(4) potential difference between the plates,

(5) capacitance with dielectric.

Solution:

Capacitance before the dielectric is introduced,

$$= C_o = \frac{\epsilon_0 A}{d} = 8.9 \times 10^{-12} \times 10^{-2}/10^{-2} = 8.9\ \mu\mu F$$

(1) Free charge $q = C_o V_o = 8.9 \times 10^{-12}$ Farad $\times$ 100 volt

$$= 8.9 \times 10^{-10} \text{ coul}$$

(2) Electric field strength in the gap. Using the relation,

$$E_o = \frac{Q}{\epsilon_0 A}$$

$$= \frac{8.9 \times 10^{-10}}{8.9 \times 10^{-12} \times 10^{-2}} = 1 \times 10^4 \text{ volt / metre}$$

(3) Electric field in the dielectric,

$$E = \frac{E_o}{K},$$

$$= \frac{1 \times 10^4}{7} = 0.14 \times 10^4 \text{ volts/metre}$$

(4) Potential Difference between the plates

$$V = E_o (d - t) + Et$$

Substituting E and E_o from above,

$$V = 1 \times 10^4 (0.5 \times 10^{-2}) + 0.14 \times 10^4 (0.5 \times 10^{-2})$$
$$= 57 \text{ volts}$$

(5) Capacitance with slab

$$\frac{Q}{A} = \frac{8.9 \times 10^{-12} \text{ coul}}{57 \text{ volt}} = 16 \; \mu\mu F$$

Example 7:

Boundary between two dielectrics. Let two isotropic dielectrics media 1 and 2 be separated by a charge free plane boundary. Let permitivities be $\in_1$ and $\in_2$ then show that,

$$\frac{\tan\alpha_1}{\tan\alpha_2} = \frac{K_1}{K_2}$$

where K_1 and K_2 are the dielectric constants of the two medium.

Solution:

As given media 1 and 2 are separated by a charge free plane boundary, *i.e.*, surface density of charge on boundary is zero, then boundary relations are:

$D_{1n} = D_{2n}$ and $E_{1t} = E_{2t}$ as discussed above

D_{1n} *i.e.*, normal component of $D_{1n} = D_1 \cos x_1$

and $\quad D_{2n} = D_2 \text{ Cos } \alpha_2$

while, $\quad E_{1t} = E_1 \sin \alpha_1$ and $E_{2t} = E_2 \sin \alpha_2$

therefore, $\quad D_1 \cos \alpha_1 = D_2 \cos \alpha_2$, (since $D_{1n} = D_{2n}$)

$E_1 \sin \alpha_1 = E_2 \sin \alpha_2$, (since $E_{1t} = E_{2t}$)

Ratio of these,

$$\frac{D_1 \text{ Cos } \alpha_1}{E_1 \text{ Sin } \alpha_1} = \frac{D_2 \text{ Cos } \alpha_2}{E_2 \text{ Sin } \alpha_2}$$

But $\quad D_1 = \in_1 E_1$,

and $\quad D_2 = \in_2 E_2$

$$\frac{\in_1 E_1 \text{ Cos } \alpha_1}{E_1 \text{ Sin } \alpha_1} = \frac{\in_2 E_2 \text{ Cos } \alpha_2}{E_2 \text{ Sin } \alpha_2}$$

or $$\frac{\epsilon_1}{\tan\alpha_1} = \frac{\epsilon_2}{\tan\alpha_2}$$

or $$\frac{\tan\alpha_1}{\tan\alpha_2} = \frac{\epsilon_1}{\epsilon_2}$$

It K_1 and K_2 are the relative permittivities of two medium, then,

$$\frac{\tan\alpha_1}{\tan\alpha_2} = \frac{K_1/\epsilon_o}{K_2/\epsilon_0} = \frac{K_1}{K_2}$$

Example 8:

Using data of example 4.6 calculate $\vec{E}$, $\vec{D}$ *and* $\vec{P}$ *then verify the equation,*

$$\vec{D} = \epsilon_o \vec{E} + \vec{P}$$

is correct both in the gap and in the dielectric.

Solution:

In the dielectric

$\vec{D} = K\epsilon_o \vec{E}$, putting the value of E and K from example 4.6.

$$\vec{D} = 7 \times 8.9 \times 10^{-12} \text{ couls}^2/\text{Newton. meter}^2 \times 1.4 \times 10^3 \text{ volts/meter}$$

$$= 8.9 \times 10^{-8} \text{ couls/meter}^2, \text{ volt} = \text{Newtons. meter}$$

and $$\vec{P} = \epsilon_o (K - 1)\vec{E}$$

$$= 8.9 \times 10^{-12} (7 - 1)(1.43 \times 10^3)$$

$$= 7.5 \times 10^{-8} \text{ coul/meter}^2$$

The D_o in air gap is given by,

$$D_o = \epsilon_o E_o \qquad \text{as } K = 1 \text{ for air}$$

$$= 8.9 \times 10^{-12} \text{ K } 1 \times 10^4$$

$$= 8.9 \times 10^{-8} \text{ coul/meter}^2;$$

In the air gap K = 1, hence,

$$P_o = \epsilon_o (K - 1)E_o = 0$$

in the dielectric,

$$\vec{D} = \epsilon_o \vec{E} + \vec{P} = 8.9 \times 10^{-12} \times 1.4 \times 10^2 + 7.5 \times 10^{-8} \text{ coul/meter}^2$$
$$= 8.9 \times 10^{-8} \text{ coul/meter}^2$$

hence verified.

In the air gap $\vec{P} = 0$

$$D_o = \epsilon_o E_o$$
$$= 8.9 \times 10^{-12} \times 1 \times 10^4$$
$$= 8.9 \times 10^{-8} \text{ coul/meter}^2$$

We can say that the relation $\vec{D} = \epsilon_o \vec{E} + \vec{P}$ holds in general, or

$$\vec{D} = \epsilon_o \vec{E} + \vec{P}, \text{ everywhere}$$

Example 9:

In example 6.7 if we maintain a constant potential difference 100 volts after the dielectric is introduced, calculate the energy.

Solution:

The electrostatics energy without dielectric per unit volume,

$$= \frac{1}{2} \epsilon_o E_o^2 \text{ jouls/meter}^3$$

and the energy density with dielectric $= \frac{1}{2} K \epsilon_o E_o^2$

substituting, K and E_o, we get,

$$U = \frac{1}{2} \times 7 \times 8.9 \times 10^{-12} \times 1 \times 10^8 \text{ jouls/meter}^3$$
$$= 31.15 \times 10^{-4} \text{ jouls/meter}^3$$

So energy density increases by K times.

(b) *However if source of emf is withdrawn as soon as dielectric is introduced then the reduces.*

Solution:

Potential difference after dielectric is introduced

$$= V_o / K$$

where V_o is potential difference before introducing the dielectric,

$$C = KC_o,\ U = \frac{1}{2}\ CV^2 = \frac{1}{2}KC_o\left(\frac{V_o}{K}\right)^2 = \frac{1}{2}\frac{C_oV_o^{\ 2}}{K}$$

$\frac{1}{2}\ C_oV_o^{\ 2}$ *is the energy before the dielectric is introduced. Hence energy decreases by factor 1/K.*

Example 10:

The polarisability of NH_3 molecule in gaseous state, from the measurement of dielectric constant is found 2.5×10^{-39} farad. meter^2 and 2.0×10^{-39} farad-meter^2 at temperature 300° K and 400° K respectively. Calculate the contribution to the polarisability because of deformation of molecules and the contribution because of permanent dipole at these temperatures.

Solution:

Using of Langevin theory,

$$\alpha_{eff} = \alpha \text{ (due to deformation)}$$

$$+ \frac{P_o^{\ 2}}{3K_bT} \text{ (due to dipole moment } P_o\text{)}$$

On putting the values,

$$2.5 \times 10^{-39} = \alpha + \frac{P_o^{\ 2}}{3K_b.300}$$

or $$2.5 \times 10^{-39} = \alpha + \frac{\beta}{300},\ \text{where } \beta = \frac{P_o^{\ 2}}{3K_b}$$

and $$2 \times 10^{-39} = \alpha + \frac{\beta}{400}$$

Solving these two equations,

$$\beta = 6 \times 10^{-37} \text{ farad.meter}^2$$

hence, $$\alpha = 0.5 \times 10^{-39} \text{ farad.meter}^2$$

The deformational polarisability = 0.5×10^{-39} farad.meter^2 and is independent of temperature.

Due to orientational polarisation at temperature 300°K, the polarisability,

$$= \frac{\beta}{T} = \frac{6\times10^{-37}}{300} \text{ Farad.meter}^2$$

$$2 \times 10^{-39} \text{ Farad.meter}^2$$

$$\text{at temperature } 400°\text{K} = \frac{6\times10^{-37}}{400}$$

$$= 1.5 \times 10^{-39} \text{ Farad.meter}^2$$

EXERCISES

1. If half the space between two concentric conducting spheres by filled with dielectric of dielectric constant K and the rest is filled with air. Show that capacitance of the capacitor thus formed will be same as if the whole part is filled with the dielectric of dielectric constant 1/2 (1 + K).
2. Calculate the induced dipole moment per unit volume of he gas if placed in a field of 6000 volts/cm. The atomic polarisability of he is 0.21×10^{-24}cm^3 and density of he is 20.6×10^{19}atoms/cc.
3. If E denotes the electric intensity in a homogeneous and isotropic dielectric, show that the actual field acting on one of the molecules of the dielectric is

$$\vec{E}_{eff} = \vec{E} + \frac{\vec{P}}{3\epsilon_0}$$

4. Derive the Classius-Mosotti's equation. How this equation is modified by Debye ?
5. What is meant by polarisability of a molecule ? Give in brief Langevin Debye theory of dielectric polarisation.
6. Derive a relation between electric susceptibility and atomic polarisability.
7. Show that the potential due to a polarised dielectric is the same as the potential due to a surface charge density over the surface of volume which has been polarised plus that due to a volume charge density.

8. Calculate the capacity of a condenser, consisting of two spheres of radii a and b, separated by a dielectric of S.I.C.K.
9. Find an expression for the capacity per unit length of two coaxial conducting cylinders placed in air.
10. Obtain an expression for energy and force of attraction between parallel plate capacitor.
11. What are polar and non-polar molecules ? Give examples.
12. What do you mean by dielectric polarisation ? What is uniform and non-uniform polarisation ? Show that for non-uniform polarisation

    $\nabla . p = -\rho_p$

13. An atom of oxygen on being polarised produces a dipole moment of 1.5×10^{-23} stat coul $\times$ cm. If the distance of the centre of the negative charge could from the nucleus is 4×10^{-15} cm, calculate the polarisability of oxygen atom.
14. Define capacitance and capacitor. On what factors the capacitance depends ?
15. Calculate the capacity of a parallel plate condenser. What will be the capacity if the space between the plates is partially filled with a slab of thickness t and dielectric constant K ?
16. Define $\vec{D}$, $\vec{E}$ and P. Establish the relation

    (1) $\vec{D} = \in_0 \vec{E} \vec{P}$

    (2) $\nabla . \vec{D} = \rho_f$

    Symbols have usual meaning.
17. Consider a surface separating two dielectric media of permittivity $\in_1$ and $\in_2$. If the dielectric field intensity in medium 1 makes an angle θ_1 with the normal to the surface. Show that the angle θ_2 which the field intensity in median 2 makes with the normal to the surface is given by

    $\in_1 \text{Cos } \theta_1 = \in_2 \text{Cos } \theta_2$

18. Define a flux tube. How the force of attraction and repulsion between charges is explained by the idea of Maxwell tubes ?

19. Explain what you mean by electronic, ionic and dipolar polarisabilities.
20. The capacitance of a parallel plate capacitor is 400 picofarad and its plates are separated by 2 mm of air. (1) What will be the energy when it is charged to 1500 volts ? (2) What will be the potential difference with same charge if plate separation is doubled ? (3) How much energy is needed to double the distance between its plates ?
21. A one core lead sheeted cable has conductor core of 0.5 cm diameter and the lead sheet has diameter 1.5 cm. The insulating material is rubber. At what voltage will the insulation break down. Given that rubber has a dielectric strength of 400KV/cm.
22. A co-axial cable consists of a copper core of 1 mm radius within outer metal sheet of 1 cm radius separated by an insulating material of dielectric constant 5. What is the capacitance per metre length of the cable in picofarad?

4

Different Series of Currents

INTRODUCTION

Let an alternating emf $\xi = \xi_o e^{j\omega t}$ be applied across a coil of inductance L of negligible resistance in parallel with a capacitor of capacitance C. The current through the inductance lags behind the applied emf in phase by $\pi/2$,

$$I_L = \frac{\xi_o}{j\omega L} e^{j\omega t} = \frac{\xi_o}{\omega L} e^{j(\omega t - \pi/2)} \text{ as } \frac{1}{j} = -j = e^{-j\pi/2}$$

The current through the capacitor leads the applied emf in phase by $\pi/2$ and is given by Eq. (1.35),

$$I_C = \frac{\xi_o}{\frac{1}{j\omega C}} e^{j\omega t} = \frac{\xi_o}{\frac{1}{\omega C}} e^{j(\omega t + \pi/2)}$$

$(j = e^{j\pi/2})$

therefore the current in the two branches differ in phase by π. The total current I_T will be the vector sum I_L and I_C and will be equal to their numerical difference, since they are 180° out of phase. Thus,

$$I_T = I_L + I_C$$

$$= \frac{\xi_o}{j\omega L} e^{j\omega t} + j\xi_{o\omega} C e^{j\omega t}$$

$$= j\left(\varepsilon C - \frac{1}{\omega L}\right)\xi_o e^{j\omega t} \qquad \text{...(1)}$$

The phase relationship between applied voltage and current complex plane.

The total impedance of the parallel circuit is given by,

$$\frac{1}{Z_T} = \frac{1}{Z_L} + \frac{1}{Z_C} = \frac{Z_L + Z_C}{Z_L Z_C}$$

or $$Z_T = \frac{Z_L Z_C}{Z_L + Z_C}$$

$$Z_T = \frac{j\omega L \dfrac{1}{j\omega C}}{j\left(\omega L - \dfrac{1}{\omega C}\right)}$$

$$= \frac{L/C}{j\left(\omega L - \dfrac{1}{\omega C}\right)}$$

$$= \frac{L/C}{\dfrac{jL}{C}\left(\omega C - \dfrac{1}{\omega L}\right)}$$

$$= \frac{1}{j\left(\omega C - \dfrac{1}{\omega L}\right)}$$

by equation (1.57),

$$Z_T = \frac{\xi_o e^{j\omega t}}{I_T} \qquad \text{...(2)}$$

The resonance is obtained when the current in the two branches is equal. Since current is out of phase in two branches, the total current will be zero. Hence by Eq. (1), at resonance

$$\omega_r C - \frac{1}{\omega_r L} = 0$$

or $$\omega_r^2 = \frac{1}{LC}$$

The resonant frequency $f_r = \dfrac{1}{2\pi\sqrt{LC}}$

This is identical with the series resonant case.

When I_T is zero. $Z_T = \infty$

Thus, at the frequency f_r of the supply voltage, the circuit offers infinite impedance to the flow of the current so that no current is drawn from the supply. Such a circuit offering infinite impedance to the flow of current is called *parallel resonant circuit* and the particular frequency f_r is called *resonant frequency.*

The *paradox of current flowing* in the coil and the condenser while no current is being supplied by the source is explained by the fact that circuit is now oscillating and the condenser is alternately charging and discharging through the coil in synchronised manner with the impressed frequency. As the coil and condenser have no resistance, the amount of energy taken by the condenser for initial charge from the source is not dissipated. The system is analogous to a frictionless pendulum which will oscillate continuously upon receiving as initial displacement without further energy being supplied.

Further at resonance,

$$Z_T = \frac{Z_L \, Z_C}{Z_L + Z_C}$$

$Z_L + Z_C$ is the total impedance when L and C are connected in series. Let us denote it by Z_S, hence we get,

$$Z_T = \frac{j\omega_r L \dfrac{1}{j\omega_r C}}{\dfrac{j\omega_r C}{Z_S}}$$

but at resonance $\omega_r L = 1/\omega_r C$

Therefore,

$$Z_T = \frac{(\omega_r L)^2}{Z_S}$$

multiplying numerator and denominator by $\xi_o \, e^{j\omega t}$, we get

$$Z_T = \frac{(\omega_r L)^2}{\xi_o e^{j\omega t}} \cdot \frac{\xi_o e^{j\omega t}}{Z_S}$$

For given supply voltage and circuit,

$$\frac{\omega_r L}{\xi_o e^{j\omega t}}$$ is constant. Hence,

$$Z_T = \text{const. } I_S \qquad ...(3)$$

where $I_S = \frac{\xi_o e^{j\omega t}}{Z_S}$ is the current when L and C are connected in series with the same emf applied. *Equation (3) shows that the impedance variation with frequency in parallel resonance case has exactly the same shape as the current variation with frequency in series resonance case.*

Parallel Resonant Circuit when Inductance L have Some Resistance

In actual practice some resistance is always associated with inductance used. Suppose an alternating emf ξ is applied to a capacitance C and inductance L having a ~esistance R connected in parallel. *Because of resistance, the current through the inductive coil will lag behind the applied emf by an angle less than $\pi/2$ and so the phase difference between I_L and I_C will be less than 180°.* Due to this the resultant impedance of the combination will be maximum at resonance instead of being infinite because of the presence of resistance in the circuit. Clearly the problem of defining parallel resonance becomes defining parallel resonance becomes more complicated. There are two criterion for solving such problems.

1. The condition of maximum impedance and
2. The condition of unity power factor.

In order to avoid confusion parallel resonance is usually taken to mean the condition for maximum impedance is called a.:ti-resonance. *Anti in the sense that in series circuit at resonance the impedance is minimum.*

Let the combination be connected to A.C. emf given by,

$$\xi = \xi_o\, e^{j\omega t}$$

The total impedance Z_T for the combination is given by,

$$\frac{1}{Z_T} = \frac{1}{Z_L} + \frac{1}{Z_C} = \frac{1}{R + j\omega L} + j\omega C$$

$$\frac{1}{Z_T} = \frac{1 + j\omega C R - \omega^2 LC}{R + j\omega L}$$

or $$Z_T = \frac{R + j\omega L}{(1 - \omega^2 LC) + j\omega CR}$$

$$= \frac{R + j\omega L}{(1 - \omega^2 LC) + j\omega CR} \cdot \frac{(1 - \omega^2 LC) - j\omega CR}{(1 - \omega^2 LC) - j\omega CR}$$

$$= \frac{R + j\omega[L - C(\omega^2 L^2 + R^2)]}{(1 - \omega^2 LC)^2 + \omega^2 C^2 R^2}$$

$$= R_o + j\omega L_o \text{ (say)} \qquad \text{...(4)}$$

where,

$$R_o = \frac{R}{(1 - \omega^2 LC)^2 + \omega^2 C^2 R^2} \qquad \text{...(5)}$$

represents the effective or apparent resistance of the coil and,

$$L_o = \frac{L - C(\omega^2 L^2 + R^2)}{(1 - \omega^2 LC^2) + \omega^2 C^2 R^2} \qquad \text{...(6)}$$

represents the effective inductance of the coil.

The magnitude of the impedance of the combination is given by,

$$Z = |ZT| = \sqrt{R_o^2 + \omega^2 L_o^2}$$

or $$Z^2 = \omega^2 [L - C(\omega^2 L^2 + R^2)]^2$$

After simplification

$$Z^2 [(1 - \omega^2 LC)^2 + \omega^2 C^2 R^2]^2$$
$$= R^2 [(1 - \omega^2 LC)^2 + \omega^2 C^2 R^2]$$
$$+ \omega^2 L^2 [(1 - \omega^2 LC)^2 + \omega^2 C^2 R^2]$$

therefore,

$$Z = \left[\frac{R^2 + \omega^2 L^2}{(1-\omega^2 LC)^2 + \omega^2 C^2 R^2}\right]^{1/2} \quad ...(7)$$

Condition for Unity Power Factor

We know that the current in the circuit will be in phase with the applied voltage *if the impedance of the combination is purely resistive* which corresponds when the power *factor is unity.* Then by eq. (4), the apparent reactance L_0 must be zero for impedance to be purely resistive, thus from Eq. (6),

$$L - C(\omega^2 L^2 + R^2) = 0$$

or $$R^2 + \omega^2 L^2 = L/C \quad ...(8)$$

And the value of angular frequency which makes the apparent reactance zero is denoted by wr given by,

$$\omega_r = 2\pi f_r = \sqrt{\frac{1}{LC} - \frac{R^2}{L^2}} \quad ...(9)$$

or $$f_r = \frac{1}{2\pi}\sqrt{\frac{1}{LC} - \frac{R^2}{L^2}} \quad ...(10)$$

This frequency fr for which power factor is unity is called the resonant frequency for the given parallel circuit corresponding to this frequency the impedance is maxi-mum (but not infinite). The current will be minimum corresponding to this circuit will work as rejector. The variation of current with frequency in parallel resonant. Clearly the current is minimum at resonant frequency. The frequency at parallel resonance given by Eq. (10) reduces to the frequency at series resonance if R is very small so that R^2/L^2 becomes negligible as compared to 1/LC. At resonance the impedance of the circuit is obtained by putting the Eqs. (8) and (9) in (7),

$$Z_{Tr} = v\sqrt{\frac{L/C}{\left[1-\left(\frac{1}{LC}-\frac{R^2}{L^2}\right)LC\right]^2 + \left(\frac{1}{LC}-\frac{R^2}{L^2}\right)C^2R^2}}$$

$$= \sqrt{\frac{L/C}{\left[1-1+\frac{R^2C}{L}\right]^2+\left(\frac{1}{LC}-\frac{R^2}{L^2}\right)C^2R^2}}$$

$$= \sqrt{\frac{L/C}{\frac{R^4C^2}{L^2}+\frac{C^2R^2}{LC}-\frac{R^4C^2}{L^2}}}$$

$$= \sqrt{\frac{L^2}{C^2R^2}}$$

$$= \frac{L}{CR} \quad \text{...(11)}$$

Dimensionally $\left|\frac{L}{R}\right|$ = [T], time constant

and similarly [CR] = [T]. Therefore $\left[\frac{L}{CR}\right]$ = [R]

is purely resistive and is known as the dynamic resistance of the combination. Be careful to note that the *resonant frequency in parallel resonance is no longer independent of the resistance in the circuit as in the case with series resonance.*

For parallel resonance to occur,

$$\sqrt{\frac{1}{LC}-\frac{R^2}{L^2}} \text{ should be real}$$

i.e., $$\frac{1}{LC}-\frac{R^2}{L^2} > 0$$

or $$R^2 < L/C$$

or $$R < \sqrt{L/C}$$

Hence the resistance R should be kept as low as possible.

CURRENT MAGNIFICATION

The A.C. supply current at resonance is obtained by dividing the supply voltage by impedance at resonance *i.e.,*

$$I_r = \frac{\xi_o e^{j\omega t}}{Z_{TR}} \text{ substituting } Z_{TR} \text{ from}$$

$$= \frac{\xi_o e^{j\omega t}}{\frac{1}{CR}} = \frac{\xi_o CR}{L} = I_{ro} \quad \text{...(1)}$$

while the oscillating current J_{Cr} in the capacitor is given by,

$$I_{Cr} = \frac{\xi_o e^{j\omega t}}{\text{Capacitative reac tan ce}}$$

$$= \frac{\xi_o e^{j\omega t}}{1/j\omega_r C} = j\xi_o \, \xi\omega_r \, C \, e^{j\omega t}$$

$$= \xi_o \omega_r C \, e^{j(\omega t + \pi/2)} \text{ (Since } j = e^{j\pi/2})$$

$$= I_{Cro} \, e^{j(\omega t + \pi/2)} \quad \text{...(2)}$$

where $I_{Cro} = \xi_o \, \omega_r \, C$

The current magnification is defined as follows:

$$m = \frac{\text{amplitude of current across capaci tan ce}}{\text{amplitude of sup ply current}}$$

$$= \frac{I_{Cro}}{I_{ro}} = \frac{\xi_o \omega_r C}{\frac{\xi_o CR}{L}}$$

$$= \frac{\omega_r L}{R} = Q \text{ (Quality factor} \quad \text{...(3)}$$

Similarly the oscillatory current I_{Lr} in the inductor is given by,

$$I_{Lr} = \frac{\xi_o e^{j\omega t}}{\text{Inductive reac tan ce}}$$

$$= \frac{\xi_o e^{j\omega t}}{j\omega_r L}$$

$$= -\frac{j\xi_o e^{j\varepsilon t}}{\omega_r L}$$

$$= \frac{\xi_o}{\omega_r L} e^{j(\omega t - \pi/2)} \text{ (as } -j = e^{-j\pi/2})$$

$$= I_{Lro} e^{j(\omega t - \pi/2)}$$

Magnification,

$$m = \frac{I_{Lro}}{I_{ro}}$$

$$= \frac{\xi_o}{\omega_r L} / \frac{\xi_o CR}{L}$$

$$= \frac{1}{\omega_r CR} = Q \qquad ...(4)$$

Thus a parallel circuit (*i.e., rejector circuit), gives a current magnification equal to the quality factor at resonance similar to the voltage magnification by a series (i.e., acceptor circuit)*

Selectivity of a Parallel Resonant Circuit

We have interpreted and concluded that impedance variation with frequency in parallel resonance case has exactly the same shape as the current variation with frequency in series resonance case. At resonance the impedance, *i.e.,*

$$Z_{Tr} = L/CR$$

By *selectivity of the circuit we mean its ability to present a high impedance at resonant frequency and much lower impedance at other frequencies.* Thus, greater the sharpness (or peakness) of the curve (*i.e.,* high Q) the greater is the selectivity. The impedance variation of the parallel circuit.

$$\frac{1}{Z_T} = \frac{1}{Z_L} + \frac{1}{Z_C}$$

$$= \frac{1}{R + j\omega L} + j\omega C$$

$$= \frac{R + j\omega CR - \omega^2 LC}{R + j\omega L}$$

or $$Z_T = \frac{R + j\omega L}{(1 - \omega^2 LC + j\omega CR}$$

In practice $R \to 0$ and we have,

$$Z_T = \frac{j\omega L}{(1 - \omega^2 LC) + j\omega CR}$$

And its magnitude directly

$$Z_T = \frac{\omega_r L}{\sqrt{(1 - \omega^2 LC)^2 + \omega^2 C^2 R^2}}$$

at $\omega = \omega_r$, $Z_T = Z_{Tr}$

Now if $\omega = \omega_r \pm \nabla\omega$, the terms ωL and $\omega^2 C^2 R^2$ will not differ appreciably from $\omega_r L$ and $\omega_r^2 C^2 R^2$ but the term $(1 - \omega^2 LC)$ will give a considerable change.

Thus when $\omega = \omega_r$ and $R \to 0$, we get,

$$Z_T = \frac{\omega_r L}{\sqrt{(1 - \omega^2 LC)^2 + \omega_r^2 C^2 R^2}}$$

It at $\omega = \omega_r + \nabla\omega$

the impedance Z_T is $L/\sqrt{2}$ times its maximum value. Thus,

$$Z_T = \frac{1}{\sqrt{2}} Z_T = \frac{1}{\sqrt{2}} \frac{L}{CR}$$

At resonance and under condition $R \to O$,

$$\omega_r = 1/\sqrt{LC}$$

$$\frac{1}{\sqrt{2}} \frac{L}{CR} = \frac{\omega_r L}{\sqrt{[1 - (\omega_r + \nabla\omega)^2 LC]^2 + \omega_r^2 C^2 R^2}}$$

On squaring, we get,

$$[1 - (\omega_r + \nabla\omega)^2 LC]^2 + \omega_r^2 C^2 R^2 = 2\omega_r^2 C^2 R^2$$

or $$[1 - (\omega_r + \nabla\omega)^2 LC]^2 = \omega_r^2 C^2 R^2$$

or $$[1 - (\omega_r^2 + \nabla\omega^2 + 2\omega_r \nabla\omega) LC]^2 = \omega_r^2 C^2 R^2$$

The term $\nabla\omega^2$ is very small and neglected,

$$[1 - \omega_r^2 LC - 2\omega_r \nabla\omega LC]^2 = w_r^2 C^2 R^2$$

$$\text{but } \omega_r^2 = 1/LC \left(\frac{2\nabla\omega}{\omega_r}\right)^2 = \omega_r^2 C^2 R^2$$

$$\text{or} \quad = \frac{2\nabla\omega}{\omega_r} = -\omega_r CR$$

$$\text{or} \quad = \frac{\omega_r}{2\nabla\omega} = \frac{1}{\omega_r CR} = Q$$

$$= \frac{\omega_r L}{R} = \frac{1}{R}\sqrt{L/C} \qquad \text{...(5)}$$

$2\nabla\omega$ is the band width of the resonance curve. Thus greater the value of Q corresponds to smaller band width the sharper is resonance and greater selectivity.

Comparative Study of a Series Resonant and a Parallel Resonant

It is of interest to compare the two circuit. The Table below gives the result discussed in the last two articles.

Series Resonant Circuit

1. Series resonant frequency is

$$f_r = \frac{1}{2\pi\sqrt{LC}}$$

2. At resonance the power factor is unity and the impedance is purely resistive

$$Z_r = R$$

3. At resonance, the impedance of the circuit is minimum while the admittance is maximum.
4. Series circuit admits maximum current at resonant frequency.
5. Series circuit is called acceptor circuit because it accepts a particular frequency and rejects others.
6. At resonance the circuit exhibits a voltage magnification equal to the quality factor.

Parallel Resonant Circuit

Here, $$f_r = \frac{1}{2\pi}\sqrt{\frac{1}{LC} - \frac{R^2}{L^2}}$$

In this case the power factor is also unity at resonance frequency but the impedance is given by,

$$Z_r = \frac{L}{CR}$$

Here at resonance the impedance of the circuit is *maximum* while the admittance is minimum. On the other hand this circuit allows a *minimum current* at the resonant frequency. This circuit is called a *rejector* because it reject only one frequency and accepts others. This circuit shows a similar *current magnification* equal to the quality factor.

Power in A.C. Circuit

Power is the rate of doing work. As in an A.C. circuit both the applied emf and the current vary continuously with time, hence the power in an A.C. circuit is equal to the product of the instantaneous emf and instantaneous current averaged over a complete cycle. The power cannot be calculated by complex number analysis *i.e.*, by directly complex ξ and complex I because the real or imaginary parts of ξI is not equal to the product of the real of imaginary parts of ξ and I respectively. Therefore the power is calculated by actual emf and actual current that is by taking imaginary parts of the complex quantities.

PURELY RESISTANCE CIRCUIT

In such circuit the current and the applied emf are in the same phase and can be expressed as,

$$\xi = \xi_o \sin \omega t; \quad I = I, \sin \omega t$$

The power $= \xi I = \xi_o I_o \sin^2 \omega t$

Average rate of doing work or power over one cycle,

$$= \frac{\xi_o I_o \int_0^{2\pi} \sin^2 \propto dx}{\int_0^{2\pi} d \propto}. \text{ where } \propto = \omega t$$

$$= \frac{\xi_o I_o}{2\pi} \frac{1}{2} \int_0^{2\pi} (1 - \cos 2 \propto) d \propto$$

$$= \frac{\xi_o I_o}{2} = \frac{\xi_o}{\sqrt{2}} \cdot \frac{I_o}{\sqrt{2}} = \xi_{rms} \times I_{rms} \qquad ...(1)$$

Unit of Power

When emf is expressed in volt and current in ampere then unit of power is watt.

POWER IN INDUCTIVE CIRCUIT

As we have discussed that the current through a pure inductance lags behind the applied emf in phase by $\pi/2$, therefore,

$$\xi = \xi_o \sin \omega t$$

and $I = I_o \sin (\omega t - \pi/2)$

Instantaneous power = xI

$$= \xi_o I_o \sin \omega t \sin (\omega t - \pi/2)$$

$$= - \xi_o I_o \sin \omega t \cos \omega t$$

$$= - \frac{\xi_o I_o}{2} \sin 2 \omega t$$

Average power over one cycle,

$$= - \frac{\xi_o I_o}{2} \frac{\int_O^{2\pi} \sin 2 \propto d \propto}{\int_O^{2\pi} d \propto}$$

$$= \frac{\xi_o I_o}{4\pi} \frac{1}{2} |\cos 2 \propto|_0^{2\pi} = 0$$

Thus power dissipation in a pure inductance is zero and the current in such circuit is known as idle or *wattles current.* The choke coil works on this principle.

PURELY CAPACITATIVE CIRCUIT

As we have discussed that the current in a purely capacitative circuit leads the applied emf in phase by $\pi/2$. Thus,

$$\xi = \xi_o \sin \omega t$$

then $I = I_o \sin(\omega t + \pi/2)$

Instantaneous power $= \xi I = \xi_o I_o \sin \omega t \cos \omega t$

$$= \frac{\xi_o I_o}{2} \sin 2\omega t$$

As the average of sin 2 ωt over one cycle is zero. Hence, the *average power is zero.* Thus the average power dissipation in a pure capacitative circuit is zero, hence the current through it is also *wattles or idle.*

CIRCUIT CONTAINING L, C AND R

As we have discussed that the emf and current are not in phase. These can be written as,

$$\xi\omega = \xi_o \sin \omega t$$

and $I = I_o \sin(\omega t + \theta)$

where phase angle θ is given by,

$$\theta = \tan^{-1} \frac{\omega L - 1/\omega C}{R}$$

Power at any instant

$$= \xi I = \xi_o I_o \sin \omega t \sin(\omega t + \theta)$$

$$= \xi_o I_o \sin \propto \sin(\propto + \theta), \propto = \omega t$$

$$= \xi_o I_o [\sin^2 \propto \cos\theta + \sin \propto \cos \propto \sin\theta]$$

$$= \xi_o I_o \left[\sin^2 \propto \cos\theta + \frac{1}{2} \sin 2 \propto \sin\theta\right]$$

The average rate of doing work or power over one cycle

$$= \frac{\xi_o I_o \cos\theta \int_0^{2\pi} \sin^2 \propto d \propto}{\int_0^{2\pi} d \propto} + \frac{\frac{1}{2}\xi_o I_o \sin\theta \int_0^{2\pi} \sin 2 \propto d \propto}{\int_0^{2\pi} d \propto}$$

As we have evaluated above that average of $\sin^2\propto = 1/2$ and that of $\sin 2\propto$ is zero. Thus, the second part does not contribute to the power is wattles, hence

Average power over one cycle

$$= \frac{1}{2}\,\xi_o\, I_o \cos\theta$$

$$= \frac{\xi_o}{\sqrt{2}} \cdot \frac{I_o}{\sqrt{2}} \cdot \cos\theta$$

$$= \xi_{rms} \cdot I_{rms} \cos\theta \quad ...(1)$$

The term ξ_{rms}. Irms is called apparent power and $\cos\theta$ is called power factor. Thus,

$$\text{power factor } (\cos\theta) = \frac{\text{True power}}{\text{Apparent power}} \quad ...(2)$$

(i) Power factor in L – C – R circuit:

$$\text{Cos } q = \frac{R}{\sqrt{R^2 + \left(\omega L - \frac{1}{\omega C}\right)^2}} \quad ...(3)$$

(ii) Power factor in inductive circuit:

$$\text{Cos } \theta = \frac{R}{\sqrt{R^2 + \omega^2 L^2}} \quad ...(4)$$

(iii) Power factor in capacitative circuit:

$$\text{Cos } \theta = \frac{R}{\sqrt{R^2 + 1/\omega^2 C^2}} \quad ...(5)$$

CHOKE COIL

As we have discussed the current lags in phase behind $\pi/2$ in a pure inductor leads in Phase by $\pi/2$ in a capacitor when connected in an A.C. circuit. Thus, there is no power loss, as the current through them is *wattless*. Hence, current in an A.C. circuit is reduced by the inclusion of a capacitor or inductor or both with- out much loss of energy. *The inductor which is used for reducing the current in the A.C. circuit is called a choke or choking coil.* A choke coil consists of thick copper wire wound closely in a large number of turns over a soft iron laminated core. The resistance of the choke coil is almost negligible, since loss of power is equal to I^2R. Thus, the loss of power is least, though there is some loss of power due to hysteresis which is much less than P R.

Further, the average power dissipated in the choke coil is given by,

$$P = \frac{1}{2} \xi_o I_o \cos \theta$$

where the power factor,

$$\cos \theta = \frac{R}{\sqrt{R^2 + \omega^2 L^2}}$$

In choke coil, L is large while R is very small. Hence cos θ is nearly zero. Therefore, the power consumed by the coil is extremely small.

Preference of a Choice Coil Over a Resistance

If resistance is used in place of inductance, the power loss is I^2R. Hence, to avoid this loss, we prefer choke coil.

ALTERNATING ELECTRIC CURRENT

An alternating current, in general, is a current that is a periodic function of time. In other words, an *alternating current is one which passes through a cycle of changes at regular intervals.* Each cycle consists of two half cycles during one of which the current is entirely positive whereas during the following half-cycle the current is entirely negative. In an alternating current circuit we are concerned with a steady state current and voltage which are oscillating sinusoidally without change in amplitude.

The variation of emf is written as,

$\xi = \xi_o \sin \omega t$ and the corresponding current is

$$I = I_o \sin \omega t$$

where ξ_o is the amplitude or peak value of voltage ξ represents the voltage at any time t and $\omega = 2\pi f$ is the angular frequency of the voltage applied. The periodic time of the oscillation is $T = 2\pi/\omega$. The term ωt is called the phase angle. The emf and the current first rise to maximum in one direction and fall to zero. The direction then quickly reverses so that emf and the current rise to maximum in the reverse direction and again fall to zero. This completes one cycle of A. C. voltage or current. *The number of such cycles per second is called the frequency (f).*

Average or Mean Value of Alternating Current or Voltage Wave over one Complete Cycle

The value of current at any instant t is given by,

$$I = I_0 \sin \omega t$$

The average value of a sinusoidal wave over one complete cycle is given by,

$$I_{av} = \frac{\int_0^T I_0 \sin \omega t \, dt}{\int_0^T dt} = -\frac{I_0}{\omega} \frac{\left|\cos \omega t\right|_C^T}{T}$$

$$= -\frac{I_0}{\omega T}\left|\cos\frac{2\pi t}{T}\right|_0^T, \text{ (since } \omega = 2\pi/T)$$

$$= -\frac{I_0}{\omega T}[\cos 2\pi - \cos 0] = -\frac{I_0}{\omega T}[1-1] = 0. \qquad ...(1)$$

Thus, *the average value of alternating current over one complete cycle is zero. The same is also true for alternating emf.*

Average Value of Alternating Current or Voltage During Half Cycle

However, the average over one half-cycle is finite. It can be obtained very easily as follows:

$$I_{av} = \int_0^{T/2} \frac{I_0 \sin \omega t \, dt}{\int_0^{T/2} dt}$$

$$= -\frac{I_0}{\omega \frac{T}{2}}[\cos \omega t]_0^{T/2}$$

$$= -\frac{I_o}{\frac{2\pi}{T}\cdot\frac{T}{2}}\left[\cos\frac{2\pi}{T}t\right]_0^{T/2}$$

$$= -\frac{I_o}{\pi}[\cos\pi - \cos 0] \qquad ...(2)$$

$$= 0.637\ I_o \qquad ...(3)$$

Similarly, for sinusoidal alternative voltage, average value over one half cycle is,

$$\xi_{av} = \frac{2\xi_o}{\pi} = 0.637\ \xi_o \qquad ...(4)$$

The average value during the second half cycle is given by,

$$I_{av} = \frac{\int_{T/2}^{T} I_o \sin\omega t\, dt}{\int_{T/2}^{T} dt}$$

After evaluating the above equation,

$$I_{av} = -\frac{2I_o}{\pi} = -\ 0.637\ I_o \qquad ...(5)$$

The average value of alternating current (or voltage) during the first and second half cycles are equal but opposite in sign i.e., they are alternately positive and negative so that the average over one cycle is zero.

VIRTUAL OR EFFECTIVE VALUE OF SINUSOIDAL VOLTAGE

The effective or virtual value of an alternating current is that steady current which, when passed through a given resistance for a certain time will develop the same amount of heat as the as the actual alternating current shall develop when passed for the same time.

The rate of heat generation in a current carrying resistance is given by I^2T. As we are interested in the average or mean value of this quantity, which we shall write as $\overline{I^2}$ where $\overline{I^2}$ represents the mean value of I^2. Thus,

$$I_{eff}^2 = \overline{I^2} \quad \text{or} \quad I_{eff} = \sqrt{\overline{I^2}} \qquad ...(6)$$

Thus, the effective value of an alternating current is the square root of the mean of the squares of the instantaneous values over one complete cycle. Due to this reason the effective current is also called the root main square (RMS) current. Let us calculate the average of I^2. If the alternating current is represented by,

$$I = I_o \sin \omega t$$

$$\overline{I^2} = \frac{\int_0^T I^2 dt}{\int_0^T dt} = \frac{\int_0^T I_o^2 \sin^2 \omega t dt}{T}$$

$$= \frac{I_o^2}{2T} \int_0^T (1 - \cos 2\omega t)\, dt$$

$$= \frac{I_o^2}{2T} \left[t - \frac{\sin 2\omega t}{2\omega} \right]_0^T$$

$$= \frac{I_o^2}{2}$$

$$\text{Hence, } I_{eff} = \sqrt{\frac{I_o^2}{2}} = \frac{I_o}{\sqrt{2}} = 0.707\, I_o \qquad ...(7)$$

Thus, the effective or RMS value of an alternating current is $1/\sqrt{2}$ times its peak value or maximum value. Similarly, the effective or virtual value of alternating emf is given by,

$$\xi_{eff} = \xi_o/\sqrt{2} = 0.707\, \xi_o \qquad ...(8)$$

Form Factor

It is defined as the ratio of the virtual or R.M.S. value to the average value of an alternating current or voltage. Thus,

$$\text{Form factor} = \frac{I_{rms}}{I_{av}} = \frac{\xi_{ms}}{\xi_{av}} \qquad ...(9)$$

Substituting values from eqs. (2) and (7), we get

$$\frac{I_o}{\sqrt{2}} \frac{\pi}{2I_o} = \frac{\pi}{2\sqrt{2}} = 1.11 \qquad ...(10)$$

The form factor gives an indication of the wave shape of an alternating voltage or current.

A more general form of alternating current or voltage may be written as,

$$\xi = \xi_o \sin(\omega t + \propto) \qquad ...(11)$$

and the current $I = I_o \sin(\propto t + \propto)$

where $\propto$ is called phase angle and added to the phase ωt. This addition does not affect the virtual or effective value (ξ_{eff}). ξ_{eff} is still given by eq. (8) which you can verify.

A.C. Circuits

We considered the transient currents in a circuit after a switch is opened or closed. The problem of determining the steady state condition was relatively simple since the circuits were energized by direct current sources (batteries). The currents after a long time were either zero or else could easily be calculated by means of Ohm's law.

An interesting and perhaps more important problem is to determine what happens when an alternating current generator is used in place of battery. Just as in the direct current case, we have to solve the differential equation for a particular circuit. The solution of the differential equation for a particular circuit may be written as the sum of complimentary function and a particular solution.

The complementary function which represents the transient solution has the same form when an alternating current source is used as when a battery is used. The complementary function may be expressed in terms of exponential functions of time. For all circuits having some resistance, the exponential terms have negative real parts. Consequently, after a relatively short time the value of the transient (current or voltage) is practically zero.

The new problem in the alternating current case is to determine the steady-state solution, which is, of course, the particular solution of the differential equation. For example, the circuit alternating current generator connected to a resistance R and a capacitance C in series. Here, *the steady state current is not zero as in the case of direct current. In one cycle, this capacitor is charged, discharges, recharged oppositely, and then discharged once more.* This happens again and again. The objective of this Chapter is to determine the current in such circuit. Basically there are two methods, namely,

1. The vector method; and
2. The complex number method.

We shall develop the complex number method.

Circuit Analysis Using Complex Number Representation

The A.C. network analysis becomes *simple and convenient by the complex number representation. A combination of real and imaginary quantity is called a complex quantity.* A complex number can be written as,

$$\vec{Z} = x + jy \qquad ...(12)$$

where $\vec{Z}$ is a complex number having a real part x and an imaginary part jy, where $j = \sqrt{-1}$. Equation (12) expresses the complex number in algebraic (or rectangular form).

The complex numbers are represented in Cartesian coordinates by a vector in a complex plane by choosing real axis along the x-axis and imaginary axis along the y-axis. The complex number is completely defined by point P since the projection along the real axis of line $\vec{OP}$ gives the real part, and its projection along y-axis gives the imaginary part of the number.

The length of the line ($\vec{OP} = \vec{Z}$) represents the absolute value or modulus (magnitude) of the complex number and angle θ is the phase of this number. Thus, the vector properties of complex quantities suggest that it may be convenient to express them in terms amplitude and direction as well as in real and imaginary parts. Hence, the complex number can be expressed in the following trigonometrical (polar) form as seen from. If point P has polar coordinates (r, θ) then we have,

$$x = r\cos\theta; \quad y = r\sin\theta$$

After substituting in Eq. (1.12), we get,

$$\vec{Z} = r\cos\theta + j\,r\sin\theta$$

$$= r(\cos\theta + j\sin\theta)$$

$$= r\,e^{j\theta} \qquad ...(13)$$

where $r = \sqrt{x^2 + y^2}$ ...(14)

and $\theta = \tan^{-1}(y/x)$ ...(15)

IMPORTANT POINTS ABOUT COMPLEX NUMBERS

Equality

Two complex numbers are said to be equal if both their real parts are equal and their imaginary parts are equal. Thus if,

$$Z_1 = x_1 + jy_1$$

$$Z_2 = x_2 + j\,y_2$$

then the equality, $Z_1 = Z_2$

gives the two real relations,

$$x_1 = x_2 \text{ and } y_1 = y_2$$

Complex Conjugate

We define the complex conjugate to $Z_1 = x_1 + j\,y_1$ as,

$$\overset{*}{Z_1} = x_1 - jy_1 \qquad ...(16)$$

$\overset{*}{Z_1}$ is a complex number such that its product with Z_1 is,

$$Z_1\,\overset{*}{Z_1} = x_1^2 + y_1^2$$

Addition and Subtraction

Complex numbers are added and subtracted like vector quantities. Let the complex numbers be represented by,

$$\vec{Z}_1 = x_1 + j\,y_1$$

and $\vec{Z}_2 = x_2 + j\,y_2$

Then the resultant complex number will

$$\vec{Z}_{Ra} = \vec{Z}_1 + \vec{Z}_2 = (x_1 + x_2) + j\,(y_1 + y_2) \text{ addition} \qquad ...(17)$$

and $ZR_s = Z_1 - Z_2\ (x_1 - x_2) + (y_1 - y_2)$ subtraction ...(19)

The magnitude of the resultant complex number is given by

$$ZR_a = \sqrt{(x_1 + x_2)^2 + (y_1 + y_2)^2} \text{ addition}$$

$$ZR_8 = \sqrt{(x_1 - x_2)^2 + (y_1 - y_2)^2} \text{ substraction} \qquad ...(19)$$

and the phase by

$$\propto = \tan^{-1} \frac{y_1 + y_2}{x_1 + x_2} \text{ addition}$$

$$\text{and } \beta = \tan^{-1} \frac{y_1 - y_2}{x_1 - x_2} \text{ subtraction} \qquad ...(20)$$

We can generalise the above rule that the sum of several complex number is a complex number whose real part is the algebraic sum of the real part of the added numbers and its imaginary part is the algebraic sum of the imaginary parts of the added numbers. The same is also true for subtraction.

Multiplication and Division

Products of complex numbers are defined according to the usual algebraic rules. For example, the product of two complex numbers Z_1 and Z_2 if expressed by Z_m is given by,

$$\vec{Z}_m = \vec{Z}_1\vec{Z}_2 = (x_1 + j\,y_1)\,(x_2 + j\,y_2)$$

$$= (x_1\,x_2 - y_1\,y_2) + j\,(x_1\,y_2 + y_1\,x_2) \qquad ...(21)$$

It can be seen that sums and products of complex numbers are also complex numbers.

Division

Division can be defined as the inverse of multiplication. When complex numbers are divided, it is first necessary to eliminate the imaginary unit contained in the denominator and then to divide the complex numbers termwise.

$$\text{Thus, } \vec{Z}_d = \frac{\vec{Z}_1}{\vec{Z}_2} = \frac{x_1 + jy_1}{x_2 + jy_2} \qquad ...(22)$$

To eliminate the imaginary unit contained in the denominator, multiply the numerator and denominator by the complex conjugate of the denominator,

$$\vec{Z}_d = \frac{\vec{Z}_1}{\vec{Z}_2} \cdot \frac{\vec{Z}_2^{\,*}}{\vec{Z}_2^{\,*}} = \frac{x_1 + jy_1}{x_2 + jy_2} \cdot \frac{x_2 - jy_2}{x_2 - jy_2}$$

$$= \frac{x_1 x_2 + y_1 y_2 + j(x_2 y_1 - x_1 y_2)}{x_2^2 + y_2^2}$$

$$= \frac{x_1 x_2 + y_1 y_2}{x_2^2 + y_2^2} + j\frac{x_2 y_1 - x_1 y_2}{x_2^2 + y_2^2} \qquad ...(23)$$

However, *the complex numbers can be multiplied and divided conveniently with less effort by transforming them from algebraic form to polar form (i.e., exponential form). By equation (13),*

$$\vec{Z}_d = r_1\, e^{j\theta}{}_1 \text{ and } \vec{Z}_2 = r^2\, e^{j\theta}{}_2$$

$$\text{Then,} \qquad \vec{Z}_m = \vec{Z}_1, \vec{Z}_2 = r_1 r_2\, e^{j(\theta_1 + \theta_2)} \qquad ...(24)$$

$$\text{and} \qquad Z_d = \frac{r_1}{r_2} e^{j(\theta_1 - \theta_2)} \qquad ...(25)$$

Thus, the product of two complex numbers is a complex number whose magnitude is the product of the magnitudes of the given complex numbers and the phase is the algebraic sum of the cofactors. The division (quotient) of two complex numbers is a complex number, the magnitude of which is the quotient of the magnitudes of the dividend and divisor, while its phase is the algebraic deference of the phases of the dividend divisor. The magnitudes and phases of the two complex numbers after multiplication and division.

Complex Representation of A.C. Circuit Quantities

We know that any vector in a complex plane can be completely specified by complex number. This suggests that the A.C. network analysis can be simplified by expressing sinusoidal voltages and currents in complex number representation rather than in vector form.

Hence the sinusoidal voltage $\xi = \xi_o \sin(\omega t + \propto)$ *and current* $I = I_o \sin(\omega t + \propto)$ *as expressed in eq. (11) that is general forms are nothing but the imaginary parts of* $\xi_o e^{j(\omega t + \propto)}$ *and* $I_o e^{j(\omega t + \propto)}$ *respectively.* Thus, alternating voltage and current can e represented conveniently by complex expression. The complex representation of voltage and current.

PURE RESISTANCE CIRCUIT CONTAINING

A circuit containing a pure resistance R with an alternating voltage $\xi = \xi_o \sin \omega t$ applied across it. This voltage is the imaginary part of the complex number $\xi_o e^{j\omega t}$. Thus, in complex number representation we can write,

$$\xi = \xi_o e^{j\omega t} \qquad ...(1)$$

If I is the instantaneous current in the circuit, at any time t, then emf equation of the circuit is,

$$RI = \xi_o e^{j\omega t} \qquad ...(2)$$

or

$$I = \frac{\xi_o e^{j\omega t}}{R} \quad \text{or} \quad I = I_o e^{j\omega t} \qquad ...(3)$$

where $I_o = \xi_o/R$

is the *maximum value of the current.*

Equation (3) represents the variation of current with time t in the circuit. Comparison of Eqs. (3) and (1) shows that the current has the same form and frequency as the applied voltage. *Thus the current is in phase with applied voltage* the complex number representation of voltage and current.

Here R is the impedance of the resistance, which when multiplied by the current through the element gives potential difference across that element. *The inverse of the impedance denoted by Y is usually called admittance.*

Thus,

$$\text{Admittance, } = Y = \frac{1}{Z} = \frac{1}{R} \qquad ...(4)$$

Circuit Containing Pure Inductance only

A circuit containing an inductor of self-inductance L with an alternating voltage $\xi = \xi_o \sin \omega t$ applied across it. This voltage is the imaginary part of the complex number $\xi_o\, e^{j\omega t}$. Thus, in complex number representation, we can write,

$$\xi = \xi_o\, e^{j\omega t}$$

Let I be the instantaneous current in the circuit, Further let dI/dt be the rate of change of current at any instant. Then emf induced in the inductor is given by,

$$\xi_{ind} = -\, Ld\ I/dt$$

The negative sign indicates that the induced emf opposes the change of current. To maintain the current flow through the inductor, the applied voltage must be equal and opposite to the induced voltage. Thus, the emf equation of the circuit is,

$$\xi = -\,\xi_{ind} = L\ dI/dt$$

or $$\xi - LdI/dt = 0$$

or $$\xi_o\, e^{j\omega t} - L\ dI/dt = 0$$

or $$dI/dt = \frac{\xi_o e^{j\omega t}}{L}$$

On integration, we get,

$$\int dI = \int \frac{\xi_o e^{j\omega t}}{L} dt$$

$$I = \frac{\xi_o}{L} \cdot \frac{1}{j\omega} e^{j\omega t} \qquad ...(5)$$

Now, $\frac{1}{j} = -\, j = e^{-j\pi/2}$

[since $e^{-j\pi/2} = \cos \pi/2 - j \sin \pi/2 = 0 - j = -\, j$]

therefore,

$$I = \frac{\xi_o}{\omega L} e^{j(\omega t - \pi/2)}$$

The quantity ω L is called *inductive reactance* of the inductance and denoted by X_L. Thus

$$X_L = \omega L$$

Hence,

$$I = \frac{\xi_o}{X_L} e^{j(\omega t - \pi/2)} = I_o\, e^{j(\omega t - \pi/2)} \qquad ...(6)$$

where $I_o = \xi_o/X_L$ is the peak value of the current.

Thus, $$X_L = \omega_L = \frac{\xi_o}{I_o} \qquad ...(7)$$

Thus, the inductive reactance is obtained by dividing peak value of voltage by peak value of current. Again, $X_L = \omega_L = 2\pi f_L$ shows that the *inductive reactance is directly proportional to the frequency of the A.C. supply voltage. Its value for D.C. supply is zero.* Further, since,

$$X_L = \frac{\xi_o}{I_o} = \frac{\xi_o / \sqrt{2}}{I_o \sqrt{2}} = \frac{\xi_{eff}}{I_{eff}} \qquad ...(8)$$

Thus, the inductive reactance may be defined as the ratio of root mean square (effective or virtue) emf across the inductance to the root mean square current flowing through it. The effect of inductive reactance $X_L = \omega L$ in an A C. circuit is the same as the resistance R in D.C. circuit to reduce the current in it, *but* without loss of *power*. The current I in the inductance L is obtained by dividing voltage across it by jωL. The impedance of the inductance is given by Eq. (5).

Equation (6) *shows that the current lags behind the voltage oy π/ 2* (or quarter of cycle) complex number representation.

A.C. Circuit Containing Capacitor only

A circuit containing a capacitor of capacitance C with an alternating voltage $\xi = \xi_o \sin \omega t$ applied across it. This voltage is the imaginary part of the complex number $\xi_o\, e^{j\omega t}$. Thus, in complex number representation, we can write,

$$\xi = \xi_o\, e^{j\omega t}$$

The instantaneous charge on the capacitor at any time t is,

$$q = C\xi = C\xi_o\, e^{j\omega t} \qquad ...(9)$$

Corresponding to this charge q, the instantaneous current which is in fact the displacement current due to the variation of voltage across the capacitor plates in the circuit is given by,

$$I = \frac{dq}{dt} = C\xi_o \cdot j\omega e^{j\omega t} = \frac{\xi_o}{1/\omega C} je^{j\omega t}$$

$$= \frac{\xi_o}{1/j\omega C} e^{j\omega t} \qquad ...(10)$$

But $e^{j\pi/2} = \cos \pi/2 + j \sin \pi/2 = j$

Hence,

$$I = \frac{\xi_o}{1/\omega C} e^{j(\omega t + \pi/2)}$$

or, $$I = \frac{\xi_o}{X_C} e^{(\omega t\_\pi/2)} = I_o\, e^{j(\omega t + \pi/2)} \qquad ...(11)$$

where $X_C = 1/\omega C$ is known as *capacitative reactance* and $I_o = \xi_o/X_C$ is the *peak value of the current.*

Clearly equation (11) shows that the *current leads the voltage by π/2.* Further, the complex number representation. The current I in the circuit is obtained by dividing exf applied by impedance of the capacitor $(1/j\omega C)$ as shown in eq. (10).

Since the capacitative reactance $X_C = 1/\omega C$ is inversely proportional to the frequency of applied voltage. Hence a condenser *offers negligible reactance at high frequency and works as "by-pass capacitor" and very high reactance at low frequency and works as "blocking condenser.*

A.C. Circuit Containing Inductance and Resistance in Series

Let us consider a circuit containing a series combination of an inductance L and resistance R with an alternating emf x = xo ejwt applied across the combination. The instaneous current in the circuit is given by,

$$I = \frac{\text{Applied emf}}{\text{Vector impedance of the circuit}} \qquad ...(12)$$

In the present case, Vector impedance $Z = R + j\omega L$ where R is impedance due to resistance and $j\omega L$ is the impedance due to inductance L. Z is also known as impedance operator. However,

Impedance $|Z|$ is given by,

$$|Z| = \sqrt{R^2 + \omega^2 L^2}$$

Thus the current,

$$I = \frac{\xi_o e^{j\omega t}}{R + j\omega L} \quad ...(13)$$

We can express $\frac{1}{R + j\omega L}$ as follows:

$$\frac{I}{R + j\omega L} = \frac{R - j\omega L}{(R + j\omega L)(R - j\omega L)}$$

$$= \frac{R - j\omega L}{R + \omega^2 L^2}$$

$$= \frac{R}{R^2 + \omega^2 L^2} - \frac{j\omega L}{R^2 + \omega^2 L^2} \quad ...(14)$$

Let, $\propto = \frac{R}{R^2 + \omega^2 L^2}$

and $\beta = \frac{\omega L}{R^2 + \omega^2 L^2}$

Then, $\frac{1}{R + j\omega L} = \propto - j\beta$

Further put $\tan\theta = \frac{\omega L}{R}$,

Then $\cos\theta = \frac{R}{\sqrt{R^2 + \omega^2 L^2}}$ and

$$\sin\theta = \frac{\omega L}{\sqrt{R^2 + \omega^2 L^2}}$$

Then, $\dfrac{I}{R + j\omega L} = \propto - j\beta = \dfrac{1}{\sqrt{R^2 + \omega^2 L^2}}[\cos\theta - j\sin\theta]$

Hence eq. 1.38 becomes, $I = \dfrac{\xi_o e^{j(\omega t - \theta)}}{\sqrt{R^2 + \omega^2 L^2}} = I_o\, e^{f(\omega t - \theta)}$ ...(15)

where impedance $Z = \sqrt{R^2 + \omega^2 L^2}$

and $\qquad I_o = \dfrac{\xi_o}{\sqrt{R^2 + \omega^2 L^2}}$

Equation (15) represents the variation of current in the circuit with time. Equation (15) also indicates the *current lags behind the applied emf in phase by an angle* θ where $\theta = \tan^{-1} \omega L/R$ (or voltage) leads the current by angle θ. Represents the complex number representation. However, the voltage ξ_R across the resistance is in phase with the current while voltage ξ_L across the inductance leading the current I by an angle $\pi/2$ and the total voltage, $\xi_o = I_o \sqrt{R^2 + \omega^2 L^2}$ leading the current by an angle $\theta = \tan^{-1} \omega L/R$. The impedance

$$Z = \frac{\xi_o}{I_o} = \frac{\xi_o / \sqrt{2}}{I_o / \sqrt{2}} = \frac{\xi_{eff}}{I_{eff}}$$

Thus, impedance may also be defined as the ratio of the r.m.s. (or virtual) value of the emf applied across the L – R circuit to the r.m.s. (or virtual) value of current flowing through the circuit.

Special Cases

1. If $R \to 0$ *i.e.*, when inductive reactance ωL is very large than to ohmic resistance R which can be neglected. Then $Z = \omega L$ and $\theta = \pi/2$.

   In *practice, it occurs when the frequency is very high.*

   Then current is given by,

$$I = I_o\, e^{j(\omega t - \pi/2)}$$

   and $I_o = \xi_o/\omega L$

Thus the peak value of current *i.e.*, I_o is inversely proportional to both self inductance and frequency. Therefore, when the frequency is high, a high voltage can produce only small current.

Further since phase lag $\theta = \tan^{-1} \dfrac{\omega L}{R}$.

Hence, greater the reactance compared with the ohmic resistance the phase lag increases. Thus, the current and the emf can be brought into phase lag increases. Thus, the current and the emf can be brought into phase by introduction of a non-inductive resistance. Practically this is adopted in A.C. voltmeters and wattmeters.

2. If $\omega L \to 0$ *i.e.*, when inductance or the frequency ($f = \omega/2\pi$) is so small as to be negligible, then ωL will be very small compared to R and we get,

$$\theta \approx 0 \text{ and } Z = R$$

$$\text{and } I = I_0 e^{j\omega t} = \frac{\xi_0 e^{j\omega t}}{R} = \frac{\xi}{R}$$

CAPACITANCE IN SERIES

Fig. 1.11 (a) shows a circuit consisting of a series combination of a resistance R and a capacitor C with an alternating emf $\xi = \xi_0\, j\omega t$ applied across the combination. The instantaneous current I is given by,

$$I = \frac{\text{Applied emf}}{\text{Vector impedance of the circuit}}$$

$$= \frac{\xi_0 e^{j\omega t}}{R + \dfrac{1}{j\omega C}} = \frac{\xi_0 e^{j\omega t}}{R - \dfrac{j}{\omega C}} \qquad ...(1)$$

where R is impedance due to resistance and $1/j\omega C$ is the impedance due to capacitance C. As in the previous case, we can express,

$$\frac{I}{R - \dfrac{j}{\omega C}} = \frac{R + \dfrac{j}{\omega C}}{R^2 + \dfrac{1}{\omega^2 C^2}}$$

$$= \frac{R}{R^2 + \dfrac{1}{\omega^2 C^2}} + \frac{j/\omega C}{R^2 + \dfrac{1}{\omega^2 C^2}}$$

Put $\propto = \dfrac{R}{R^2 + \dfrac{1}{\omega^2 C^2}}$

and $\beta = \dfrac{1/\omega C}{R^2 + \dfrac{1}{\omega^2 C^2}}$

and $\tan\theta = \dfrac{1}{\omega CR}$

We get, $$\frac{I}{R - \dfrac{j}{\omega C}} = \alpha + j\beta$$

$$= \frac{1}{\sqrt{R^2 + \dfrac{1}{\omega^2 C^2}}}[\cos\theta + j\sin\theta]$$

$$= \frac{1}{\sqrt{R^2 + \dfrac{1}{\omega^2 C^2}}} e^{j\theta}$$

Hence Eq. (1) reduces to

$$I = \frac{\xi_o e^{j(\omega + \theta)}}{\sqrt{R^2 + \dfrac{1}{\omega^2 C^2}}}$$

or $$I = I\, e^{j(\omega + \theta)} \qquad \text{...(2)}$$

where, $$I_o = \frac{\xi_o}{\sqrt{R^2 + \dfrac{1}{\omega^2 C^2}}}$$

indicates the *peak value of current.* Equation (2) represents the variation of current with time and shows that *current leads the applied voltage in phase by an angle* θ.

$$\theta = \tan^{-1} \frac{1/\omega C}{R}$$

The impedance, $Z = \sqrt{R^2 + \frac{1}{\omega^2 C^2}}$

$$= \sqrt{(\text{Resistance})^2 + (\text{Capacitative reactance})^2}$$

The complex number representation of current and voltage. *The current leads the applied voltage by phase is the same as the voltage lags behind the current in phase.* The voltage ξ_R across the resistance is in phase with current, while the voltage ξ_C across the condenser lagging the current I by an angle $\pi/2$. Finally the total voltage,

$$\xi_o = I_o \sqrt{R^2 + \frac{1}{\omega^2 C^2}}$$

lagging the current I by an angle $\theta = \tan^{-1} \frac{1/\omega C}{R}$.

Special Cases

1. If $C \to 0$, then $I_0 \to 0$, this means no current will flow in the circuit with zero capacity and this causes a break in the circuit. The same result will be obtained with very low frequencies.
2. If $C \infty \infty$, then $I_0 = \xi_0/R$. Thus a condenser of infinite capacity does not offer any resistance to A.C. and $\tan \theta = 0$ or $\theta \to 0$. This means the introduction of such a condenser causes no phase change. The same result follows if the frequency is very high.
3. If $R \to 0$, then $I = \xi_0\, \omega C$ and $\tan q \to \infty$ or $\theta = \pi/2$. Thus the current is leading by $\pi/2$ in phase with respect to emf.

A.C. CIRCUIT WITH RESISTANCE, CAPACITANCE

Let a sinusoidal alternating emf $\xi = \xi_0\, e^{j\omega t}$ be applied to the series combination of L.C.R. The vector impedance of the combination is sum of three parts.

1. R due to resistance R
2. $j\omega L$ due to inductance L
3. $1/j\omega C = -j\omega C$ due to capacitance C.

Thus in vector form the emf eqn. of the circuit is

$$\left[R + j\left(\omega L - \frac{1}{\omega C}\right)\right] I = \xi_o \, e^{j\omega t} \qquad ...(1)$$

where I represents the instantaneous current in circuit, is given by,

$$I = \frac{\xi_o e^{j\omega t}}{R + j\left(\omega L - \frac{1}{\omega C}\right)} \qquad ...(2)$$

Adopting the same procedure as in the previous article, we can write,

$$\frac{1}{R + j\left(\omega L \frac{1}{\omega C}\right)} = \frac{R - j\left(\omega L - \frac{1}{\omega C}\right)}{R^2 + \left(\omega L - \frac{1}{\omega C}\right)^2}$$

$$= \propto - j\beta$$

where, $x = \dfrac{R}{R^2 + \left(\omega L - \frac{1}{\omega C}\right)^2}$

and $\beta = \dfrac{\omega L - \frac{1}{\omega C}}{R^2 + \left[\omega L - \frac{1}{\omega C}\right]^2}$

and if we put,

$$\tan\theta = \frac{\omega L - \frac{1}{\omega C}}{R}$$

Then, $\cos\theta = \dfrac{R}{\sqrt{R^2 + \left(\omega L - \frac{1}{\omega C}\right)^2}}$

and $\sin\theta = \dfrac{\omega L - 1/\omega C}{\sqrt{R^2 + \left(\omega L - \frac{1}{\omega C}\right)^2}}$

Hence,

$$\frac{1}{R + j\left(\omega L - \frac{1}{\omega C}\right)} = \frac{1}{\sqrt{R^2 + \left(\omega L - \frac{1}{\omega C}\right)^2}}(\cos\theta - j\sin\theta)$$

$$= \frac{1}{\sqrt{R^2 + \left(\omega L - \frac{1}{\omega C}\right)}} e^{-j\theta}$$

Then the expression for current *i.e.,* eq. (3) reduces,

$$I = \frac{\xi_o}{\sqrt{R^2 + \left(\omega L - \frac{1}{\varepsilon C}\right)^2}} e^{j(\omega t - \theta)}$$

or $$I = I_o \, e^{j(\omega t - \theta)} \qquad ...(3)$$

where, $$I_o = \frac{\xi_o}{\sqrt{R^2 + \left(\omega L - \frac{1}{\omega C}\right)^2}} \qquad ...(4)$$

I_o is the *peak value of the current.* Ans.

$$\sqrt{R^2 + \left(\omega L - \frac{1}{\omega C}\right)^2} = Z$$

where Z is called the *impedance* of the circuit (in magnitude), which is the ratio of peak emf to the peak current. Again impedance, from eq. (4), is given by

$$Z = \frac{\xi_o}{I_o} = \frac{\xi_o / \sqrt{2}}{I_o \sqrt{2}} = \frac{\xi_{eff}}{\xi_{eff}}$$

Thus, the impedance of a L.C.R. circuit can be defined as the ratio of the r.m.s. value of emf applied to the r.m.s. current flowing through it. The *current lags the applied voltage in phase by an angle* θ *given by,*

$$\theta = \tan^{-1} \frac{\omega L - \frac{1}{\omega C}}{R}$$

The variation of current and voltage with time. In other words, the emf and current have a phase difference of θ.

This phase difference in the complex plane. The voltage across the resistance ξ_R is shown in phase with the current while, which is the net voltage across the combination of L and C and leads the current by an angle $\pi/2$. The total voltage ξ_o is shown leading the current by an angle,

$$\theta = \tan^{-1} \frac{X_L - X_C}{R}$$

Special Cases

The phase difference (*i.e.*, leading or lagging) in fact depends on the value of θ which depends on value of X_L and X_C. Depending on the values of X_L and X_C there are three important cases.

1. If $\omega L > \frac{1}{\omega L}$ i.e., when impressed frequency ($f = \omega/2\pi$) is very large, the phase angle θ will be positive and the current will lag behind the applied emf. The potential difference $I_o X_L = I_o \omega_L$ across the inductance is greater than potential difference $I_o/\omega C$ across the capacitance and the circuit behaves as an inductive circuit. The vector voltage in complex plane.

2. If $\omega L < \frac{1}{\omega C}$, the phase angle θ will be negative and the current will lead the applied emf. The circuit behaves as capacitative circuit. The vector diagram for the case in complex plane.

3. If $\omega L = 1/\omega C$ a very important and interesting case arises. In this case,

$$\theta = \tan^{-1} \frac{\omega L - \frac{1}{\omega C}}{R} = 0$$

i.e., the phase angle θ becomes zero and the emf and current will be in phase. *The potential difference across the inductance and capacitance are equal in magnitude but opposite in phase and therefore cancel out* and the whole voltage is dropped across the resistance. Therefore, the circuit behaves as *pulley resistive circuit.* The vector diagram in complex plane. Under this condition, the peak current will be maximum given by I_o max

$$I_{0max} = \frac{\xi_0}{\sqrt{R^2 + \left(\omega L - \frac{1}{\omega C}\right)^2}} = \frac{\xi_0}{R} \qquad ...(5)$$

Thus, in this particular case when $\omega L = 1/\omega C$, the emf and current will be in phase and *current is maximum*. The circuit is said to be a *series resonant circuit* and the phenomenon of maximum current is called *resonance*. Thus at resonance,

$$\omega_r L = 1/\omega_r C$$

or $$\omega_r = 1/\sqrt{LC}$$

or $$f_r = \frac{1}{2\pi\sqrt{LC}} \qquad ...(6)$$

Here subscript r corresponds to resonance.

The frequency at which resonance occurs is known as *resonance frequency.* Evidently from Eq (6) the resonant frequency is the same as the frequency of the oscillatory discharge of a series LCR circuit when the resistance is low. Thus, the series LCR circuit is in resonance with applied voltage. The *frequency of applied voltage coincides with the natural frequency of the circuit.* Gives the vector diagram of voltage and current for this circuit on the complex plane under this condition the variation of peak current with frequency for resonant circuit.

Sharpness of Resonance of a Series LCR Circuit

We have discussed in the previous article that the current amplitude is,

$$I_o = \frac{\xi_0}{\sqrt{R^2 + \left(\omega L - \frac{1}{\omega C}\right)^2}}$$

At resonance $\omega L = 1/\omega C$

therefore $I_{omax} = \xi_o/R$

Let us examine the effect of circuit resistance on the current when the supply frequency is varied continuously from a value below the resonant frequency to above it, when the resistance of the circuit is (1) low, (2) medium and (3) high. The curves obtained by plotting amplitude of current (or voltage drop across a fixed resistance) against ω (or f) are known as *resonance curves*. The variation of current amplitude as function of frequency for three different values of resistance R.

Clearly, as the resistance in the circuit is reduced the resonance curve becomes sharper. The peak value of I_o shown that the *circuit responds only to the frequency exactly equal to its natural frequency of the circuit* ω_r ($\omega_r = 1/\sqrt{LC}$ for low resistance in LCR circuit) and to none others. The resonance here is, therefore, said to be *sharp*.

Thus, *sharpness of resonance is, in a way, a measure of the* rate of *fall of amplitude from its maximum value at resonant frequency on either side of it.* The more quickly the current amplitude falls for changes of frequency f_r the sharper is said to be resonance.

On the other hand, if the amplitude remains more or less at it peak value over an appreciable range of frequency on either side o $\omega_r = \frac{1}{\sqrt{LC}}$ resonant frequency *i.e.*, even when $\omega_r \neq \frac{1}{\sqrt{LC}}$. Thus, the *circuit responds to a number of frequencies* near about f_r on either side of it. The resonance in this case, is therefore, said to be *flat*.

Variation of Voltages Across Inductance and Capacitance is a Series L.C.R. Resonant Circuit

The voltage drop across the inductance coil is

$$\xi_L = I_o X_L = \frac{\xi_o}{Z} X_L = \frac{\xi_o}{\sqrt{R^2 + \omega^2 L^2}} \omega L$$

For frequency considerably above resonant frequency $\omega L > R$, we have,

$$\xi_L = \frac{\xi_o \omega L}{\omega L} = \xi_o$$

equal to *amplitude of impressed voltage.*

Similarly, the voltage drop across the capacitor is,

$$\xi_C = I_o X_C = \frac{\xi_o}{\sqrt{R^2 + \frac{1}{\omega^2 C^2}}} - \frac{1}{\omega C}$$

For very low frequencies approaching zero, R^2 can be neglected in comparison with $1/\omega^2 C^2$, we have,

$$\xi_C = \frac{\xi_o}{1/\omega C} - \frac{1}{\omega C} = \xi_o$$

equal to the impressed voltage.

The variation of ξ_L and ξ_C with ω. Both of these curves are similar in shape in the vicinity of the resonance to the curves. This is because the resonance phenomenon takes place in a very limited range. The reactance across which the voltage is developed remains substantially the same in the region round about the resonance. The product of current and reactance, therefore, varies with frequency practically in the same way as current.

Q of a Circuit

Qualitatively the sharpness of resonance curve is determined by a quality factor called "Q" of the circuit. It is *defined as there ratio of reactance of either the inductance or capacitance at the resonant frequency to the total resistance of the circuit.* Thus, mathematically,

$$Q = \frac{X_L}{R} = \frac{\omega_r L}{R}$$

or

$$Q = \frac{X_C}{R} = \frac{1}{\omega_r CR} \qquad ...(7)$$

Since at resonance $X_L = X_C$, *i.e.*, $\omega_r L = 1/\omega_r C$, we get the same value of Q from both expressions.

How the sharpness depends on "Q" can be understood as follows. Since at resonance the current is inversely proportional to the resistance ($I_o = \xi_o/R$). So, directly proportional to Q (see Eq. 7). The value of the current for frequencies appreciably different form resonant frequency is smainly determined by the reactance and not by the resistance. Thus the current is nearly independent of the resistance of the circuit. The result of increasing the resistance of the circuit (*i.e.,* lowering Q) reduces the reasons at the resonant frequency but leaves the other frequencies nearly unaffected. Therefore, if Q is large the resonance curve is sharp.

"Q" is also defined in terms of *lower and upper half-power frequencies.* The half-power frequencies are the frequencies at which the power dissipation in the circuit drops to one half its resonance value. If f_1 and f_2 are the lower and upper half power frequencies and f_r is the resonant frequency,

The power at resonant frequency $= I^2_{omax}$

$$= \left(\frac{\xi_o}{R}\right)^2 R_o$$

The power dissipation at f_1 is given by,

$$I_1^2 R = \frac{1}{2}\left(\frac{\xi_o}{R}\right)^2 R$$

where I_1 is current at lower half power frequency

or $$I_1 = \frac{1}{\sqrt{2}} \cdot \frac{\xi_o}{R} = \frac{1}{\sqrt{2}} \text{ current at resonance.}$$

Similarly, I_2 the current at upper half power frequency is

$$I_2 = \frac{1}{\sqrt{2}} \frac{\xi_o}{R} = \frac{1}{\sqrt{2}} \text{ current at resonance}$$

Thus at lower or upper half power frequencies the current amplitude becomes $1/\sqrt{2}$ times the current amplitude at resonance.

Hence from eq. (5) at lower or upper half power frequency, we have,

$$\frac{\xi_o}{\sqrt{R^2 + \left(\omega L - \frac{1}{\omega L}\right)^2}} = \frac{1}{\sqrt{2}} \cdot \frac{\xi_o}{R}$$

On squaring and rearranging, we get,

$$R^2 + \left(\omega L - \frac{1}{\omega C}\right)^2 = 2R^2$$

or
$$\left(\omega L - \frac{1}{\omega L}\right)^2 = R^2$$

This equation has two solutions one for lower half and other for upper half power frequencies. Thus,

$$\omega_1 L - \frac{1}{\omega_1 C} = -R \quad ...(8)$$

$$\omega_2 L - \frac{1}{\omega_2 C} = +R \quad ...(9)$$

Adding these two equations, we get,

$$(\omega_1 + \omega_2) L - \frac{1}{C}\left(\frac{1}{\omega_1} + \frac{1}{\omega_2}\right) = 0$$

or
$$(\omega_1 + \omega_2) L - \frac{\omega_1 + \omega_2}{\omega_1 \omega_2}\frac{1}{C} = 0$$

or
$$\omega_1 \omega_2 = \frac{1}{LC} \quad ...(10)$$

or
$$\omega_1 \omega_2 = \omega_r^2 \text{ (by eq. 6)} \quad ...(11)$$

Thus, the resonant frequency f_r is the geometric mean of the upper and lower half power frequencies.

Subtracting Eq. (8) from Eq. (9), we get,

$$(\omega_2 - \omega_1) L + \frac{1}{C}\left(\frac{1}{\omega_1} - \frac{1}{\omega_2}\right) = 2R$$

$$(\omega_1 - \omega_2) L + \frac{\omega_2 - \omega_1}{\omega_1 \omega_2}\frac{1}{C} = 2R$$

Substituting $\omega_1 \omega_2$ from Eq. (10),

$$(\omega_2 - \omega_1)\left(L + \frac{1}{1/LC}\frac{1}{C}\right) = 2R$$

$$\omega_2 - \omega_1 = R/L$$

Therefore,

$$\frac{\omega_r}{\omega_2 - \omega_1} = \frac{f_r}{f_2 - f_1} = \frac{\omega_r L}{R} = Q \qquad ...(12)$$

Q in terms of *band width* is expressed as,

$$Q = \frac{\omega_r}{2\nabla\omega} \qquad ...(13)$$

where $\nabla\omega$ is the change in ω on either side of ω_r as a result of which the power dissipation becomes half its value at the resonant frequency. Thus,

$$\omega_1 = \omega_2 - \nabla\omega$$

and $\omega_2 = \omega_r + \nabla\omega$

The frequency difference,

$$\omega_2 - \omega_1 = 2\nabla\omega$$

$2\nabla\omega$ is called the *band width.*

Thus, greater value of Q corresponds to smaller $\omega_2 - \omega_1$ or $2\nabla\omega$ band width, the sharper is the resonance.

Selectivity

If the applied alternating voltage has a number of frequency components (for example, the signal received by a radio antenna has wide range of frequencies). Then a series L.C.R. circuit will give a maximum response (*i.e.,* it will pass a maximum current and have a maximum potential difference across its inductance) for only that frequency component which has the frequency,

$$f_r = \frac{1}{2\pi\sqrt{LC}}$$

Thus out of number of available frequencies it selects one frequency for which the current is maximum and for other frequencies the current is comparatively very small, in other words it shows selectivity. The resonant circuit is consequently known as the "acceptor circuit" as it practically accepts one frequency component of the applied voltage and rejects other. It may be noted that sharper the response (*i.e.*, higher the Q value) the more highly selective the acceptor circuit is.

Voltage Magnification

The circuit Q is also a measure of voltage magnification in the series L.C.R. circuit which can be understood as follows.

At resonance the potential difference across inductance and capacitance are equal and 180° out of phase and hence cancel. Therefore, the only potential difference at resonance is across the resistance.

As resonant current is maximum and given by,

$$I_{omax} = \frac{\xi_o}{R}$$

potential difference across resistance R = I_{omax} R = ξ_o

Thus, at resonance the potential difference across resistance is equal to the applied emf. The voltage magnification in L.C.R. circuit is defined as,

$$\text{Voltage magnification} = \frac{\text{p.d. across the induc tan e or capaci tan e}}{\text{Spplied voltage}}$$

$$m = \frac{I_{o\,max}\,\omega_r L}{\xi_o}$$

$$= \frac{\xi_o}{R}\frac{\omega_r L}{\xi_o} = \frac{\omega_r L}{R} = Q = \frac{1}{R}\sqrt{\frac{L}{C}} \qquad ...(14)$$

Similarly for capacitance,

$$m = \frac{I_{o\,max}}{\xi_o}\frac{1}{\omega_r C}$$

$$= \frac{\xi_o}{R\xi_o}\frac{1}{\omega_r C}$$

$$= \frac{1}{\omega_r CR} = Q = \frac{1}{R}\sqrt{\frac{L}{C}} \qquad ...(15)$$

Thus, the voltage magnification of L.C.R. circuit is equal to its quality factor Q at resonance.

SOLVED EXAMPLES

Example 1:

A circuit consists of a résistance and capacitance in series. If an alternating emf of 180 volts and 100 c/sec frequency is applied to it, calculate the value of resistance and capacitance when the maximum current is 6 amps. and active power is 360 watts.

Solution:

The circuit resistance is given by power relation

$$R = \frac{P}{I^2} = \frac{360}{6^2} = 10 \text{ ohms.}$$

and the circuit impedance is

$$Z = \frac{V}{I} = \frac{180}{6} = 30 \text{ ohms.}$$

The capacitive reactance is

$$X_c = \sqrt{Z^2 - R^2} = \sqrt{30^2 - 10^2} = 28.3 \text{ ohms.}$$

therefore, the capacitance is

$$C = \frac{1}{\omega X_c} = \frac{1}{2 \times 3.14 \times 100 \times 28.3}$$

$$= 26.2 \text{ micro farad.}$$

Example 2:

A resistance of 10 ohms is joined in series with an inductance of 0.5 henry. What capacitance must be put in series with the combination to obtain maximum current? What will be the potential drop across each element of circuit, if it is connected to 200 V 50 H_x mains?

Solution:

Let the capacity be C, then $\left(\frac{1}{\omega C} = \omega L\right)$ of maximum current which corresponds to resonance or $$C = \frac{1}{\omega^2 L} = \frac{10^6}{0.5\times(2\times3.14\times50)^2} \text{ microfarad}$$

$= 20.25$ microfarad

$$\text{Current } I = \frac{V}{R} = \frac{200}{10} = 20 \text{ ampere}$$

potential difference across R = RI = $10 \times 20 = 200$ V.

Potential difference across L or C = ωLI

$= 0.5 \times 2 \times 3.14 \times 50 \times 20 = 3142$ volts.

Example 3:

A 48 ohms resistor is connected in series with an inductance of 450 mH and a capacitance of 9 microfarad. (i) What is the resonant frequency? (ii) What will be the current in the circuit when it is supplied from a 120 V source operating at resonant frequency? (iii) What is the voltage across the inductance under these conditions?

Solution:

$\omega L = 1/\omega C$ (condition of resonant)

$$\omega = \frac{1}{\sqrt{LC}} = \frac{1}{\sqrt{450\times10^{-3}\times9\times10^{-6}}} = 2\pi f$$

$$\text{(i) } f = \frac{1}{2\pi\sqrt{450\times9\times10^{-9}}} = 79 \text{ cycles/sec.}$$

$$\text{(ii) } I_{max} = \frac{120}{48} = 2.5 \text{ amp.}$$

voltage across inductance = $\omega L\, I_{max} = 2\pi \times 79 \times 450 \times 10^{-3} \times 2.5$

$= 557.0$ volts.

Example 4:

In a series L.C.R. circuit the values of the various quantities are as follows:

$\xi = 100$ volts, $f_o = 60$ c/sec. $R = 1$ ohm,

$L = 1$ henry and $C = 7.C4$ microfarad

Prove that the voltage magnification is equal to quality factor Q.

Solution:

Inductive reactance $X_L = \omega_o L = 2\pi r_o L$

$= 2 \times 3.14 \times 60$

$= 377$ ohms (approx.)

Capacitive reactance

$$= X_C = \frac{1}{\omega_o C} = \frac{1}{2\pi f_o C} = 377 \text{ ohms (approx).}$$

At resonance maximum current,

$$I_{max} = \frac{E}{\sqrt{R^2 + (X_L - X_C)^2}} = \frac{E}{R} = 100 \text{ amp.}$$

Therefore potential difference across resonance resistance

$\xi R = I_{max}\ R = 100$ volts.

Potential difference across inductance

= potential difference across capacitance

$$= \omega_o L\ I_{max} = \frac{I_{max}}{\omega C} = 377 \times 100 = 37700 \text{ volts}$$

$$= Q\xi \left(\text{as} Q = \frac{\omega_o L}{R} = \frac{1}{\omega_o CR} = 377 \right)$$

therefore,

Voltage magnification is equal to quality factor Q.

Example 5:

An alternative valtage of 110 V 50 cycle is applied to a circuit which contains an inductance of 0.02 henry and a resistance of 10 ohms in series. Determine the current and the phase lag.

Solution:

We have

$$I_o = \frac{\xi_o}{\sqrt{R^2 + \omega^2 L^2}} = \frac{110}{\sqrt{10^2 + (2 \times 3.1416 \times 50 \times 0.02)^2}}$$

$$= 10.5 \text{ amp.}$$

$$\theta = \tan^{-1}\frac{\omega L}{R} = \tan{-1}\,\frac{2 \times \pi \times 50 \times 0.02}{10}$$

$$= \tan^{-1}\frac{\pi}{5}$$

$$= 32°8' \text{ lagging.}$$

Example 6:

A circuit having a pure inducatance of 0.01 henry in series with a resistance of 5 ohms is fed by 200 V. 50 c/sec. a.c. supply. Calculate the current and its phase, the drop of potential across the inductor and resistor, and the power.

Solution:

$$I_o = \frac{\xi_o}{\sqrt{R^2 + \omega^2 L^2}} = \frac{200}{\sqrt{25 + (2\pi f L)^2}} \text{ on substitution}$$

$$= \frac{200}{\sqrt{25 + (3.14)^2}} = \frac{200}{5.91} = 33.9 \text{ amps.}$$

The angle of lag of current is given by,

$$\theta = \tan^{-1}\frac{\omega L}{R} = \tan{-1}\left(\frac{3.14}{5}\right) = 32°$$

Now the potential difference across the resistance and the inductance are,

$$\xi_R = I_o R = 33.9 \times 5 = 169.5 \text{ volts}$$

and $\xi_L = I_o\,\omega L = 33.9 \times 3.14 = 106$ volts.

Power dissipation is

$$P = I_o^2 R = (33.9)2 \times 5 = 5746 \text{ volts.}$$

Example 7:

An electric lamp marked 100 volts. d.c. consumes a current of 10 amperes. It is connected to a 200 volts 50 cycles a.c. mains. Calculate the inducatance of the required whole.

Solution:

Given $I_o = 10$ amp. $R = \frac{V}{I} = \frac{100}{10} = 10$ ohms

$\xi_o = 200$ volts, $f = 50$ c/sec.

$\omega = 2\pi f = 2\pi \times 50 = 100\pi$

$$\text{Now } I_o = \frac{\xi_o}{\sqrt{R^2 + \omega^2 L^2}}$$

$$\text{or} \quad 10 = \frac{200}{\sqrt{(10)^2 + (100\pi L)^2}}$$

$$\text{or} \quad (100\ \pi L)^2 = 300$$

therefore,

$$L = \frac{\sqrt{300}}{100\pi} = 0.055 \text{ henry}$$

Example 8:

Find the capacity of a condenser to run a 30 volt, 10 watt lamp when connected series with an alternating e.m.f. of 220 volts and frequency 50 cylces per second.

Solution:

Here, power of the lamp = $V \times I = 10$ watts, $V = 30$ volts.

therefore.

$$I = \frac{W}{V} = \frac{10}{30} = \frac{1}{3} \text{ amp.}$$

$$\text{and } R = \frac{V}{I} = \frac{30}{1/3} = 90 \text{ ohms.}$$

$$\text{Now } I_o = \frac{\xi_o}{\sqrt{R^2 + \left(\frac{1}{\omega C}\right)^2}}$$

$$\text{or} \qquad \frac{1}{3} = \frac{220}{\sqrt{(90)^2 + \left(\frac{1}{100\pi C}\right)^2}}$$

On solving, we get, $C = 4.866 \times 10^{-6}F = 4.866\mu F$

Example 9:

An alternating emf of 200 volts 50 cycles per sec. is applied to a condenser in series with a 100 volts 50 watt lamp. Find the capacity of the condenser.

Solution:

$$\xi I = 50$$

$$I = 50/\xi = 50/100 = \frac{1}{2} \text{ amp.}$$

$$R = \frac{\xi}{I} = \frac{100}{1/2} = 200 \text{ ohms}$$

$$\text{and } R^2 + \frac{1}{\omega^2 C^2} = \left(\frac{\xi}{I}\right)^2 = \left[\frac{200}{1/2}\right]^2 = 400^2$$

$$\frac{1}{\omega^2 C^2} = 400^2 - 200^2 = 12 \times 10^4$$

$$\omega^2 C^2 = \frac{1}{12 \times 10^4}$$

$$\omega = 2\pi f = 2\pi \times 50 = 100\pi$$

$$C = \frac{1}{100\pi \times 100 \times \sqrt{12}}$$

$$= 9.16 \times 10^{-6} \text{ farads}$$

$$= 9.16 \text{ microfarads}$$

EXERCISES

1. A non-inductive resistance of 12 ohms is connected in series with a coil of negligible resistance having an inductance of 0.016 henry. An alternating emf of 130 volts 50 cycles is applied to the circuit. Calculate the current and potential difference across the resistance and the inductance.

2. A circuit contains a resistance of 4 ohms and an inductance of 0.68 henry and an alternating effective emf of 500 volts at a frequency of 120 cycles per second is applied to it. Find the value of the effective current in the circuit and power factor.

3. A RLC circuit has R = 2 ohms, L= 0.005 henry and C = 600 microfarad. If potential difference of 100 volts (rms) at/frequency 50 is applied, calculate (i) current at resonance (ii) power and (iii) drop in volts along each element of the circuit.

4. An inductance of S milli henry is connected in parallel with the condenser of capacity 500 microfarad. The two are joined in series with a non-inductive resistance of 4 ohms and a source of alternating emf of 100 volts 50 cycles. Determine/the current in the main circuit. At what frequency is the circuit anti-resonant?

5. Why is a capacitor of practically infinite d.c. resistance? Explain the passage of alternating current through a purely capacitive circuit.

   Deduce equation for current in a capacitive circuit. Plot the current and voltage curves for the capacitive circuit.

6. Explain the impedance and reactance in an a.c. circuit. Distinguish between the two.

7. Derive expressions for current and impedance when an alternating emf is applied to a circuit having (i) capacity and resistance in series (ii) inductance and resistance in series.

8. Derive an expression for the amplitude and phase of the current in a series circuit consisting of a resistance R, a self-inductance L and a capacitance C, when an alternating emf is applied.

9. Describe the behaviour of a series R—L—C circuit for the case when $X_C > X_L$ and $X_L < X_C$. Discuss the phenomenon of resonance in series R—L—C circuit.

10. An alternating emf of 200 volts and 50 periods per second is applied to a condenser in series with a 20 volt, 5 watt lamp. Find the capacity of the condenser required to run the lamp.
11. A resistance of 10 ohms is joined in series with an inductance of 0.5 henry. What capacitance should be put in series with the combination to obtain the maximum current ? What will be the potential difference across the resistance, inductance and capacity? The current is being supplied by 200 volts and 50 cycles mains.
12. Alternating emf of 100 virtual volts is applied to a circuit of resistance 0.5 ohm and inductance 0.01 henry, the frequency being 50 cycles/sec. What is the current and lag in the time between emf and current?
13. What do you understand by the quality factor (Q) of a circuit? Explain the dependence of the sharpness of the resonance on Q in series L—C—R circuit.
14. Derive an expression for Q in terms of bandwidth of the curve.
15. Prove that Q is the factor by which the applied voltage is magnified across L and C at resonance. Why LCR circuit is often called an acceptor circuit!
16. What is the condition of resonance when an inductor is shunted by a capacitor and explain the paradox of current flowing in the coil and condenser while no current is being supplied by the source at resonance ?
17. What is the difference between parallel resonance and anti-resonance? Derive an expression for impedance when an inductive coil is shunted by a condenser.
18. What do you understand by wattless and power components of an alternating current ?
19. Find the power in an a.c. circuit having: (i) resistance and capacity (ii) resistance and inductance. What is meant by the power factor of a circuit?

5

Applicaition of Potentiometer Current

INTRODUCTION

In potentiometer, *the accuracy increases as the length of the wire is increased. But a potentiometer of greater length becomes unwieldy.* This difficulty is overcome by using compact forms of potentiometer.

A potentiometer is a device used to measure potential difference. The device is simply to compare an unknown emf with a known one. It consists of a uniform wire AB of resistance R. A steady current is passed through the wire AB by a constant source of emf ξ through the rheostat Rh. ξ_1 is emf of the cell which is to be measured and is connected in the circuit. When the current is steady and constant, the potential gradient is constant.

Let resistance per unit length of the wire AB be ρ and the steady current passing through the wire be I. Then the potential gradient that is the fall of potential per unit length of the wire is I ρ. Connect ξ_1 by K and move the sliding contact till the galvanometer gives no deflection. If l_1 is the length of the potentiometer wire for balancing ξ_1, then

$$\xi_1 = I\rho l_1.$$

Similarly connect x^2 by K and repeat the process, now if l_2 is the length of the potentiometer wire for balancing x_2, then,

$$\xi_2 = I\rho l_2.$$

Hence ratio,

$$\frac{\xi_1}{\xi_2} = \frac{I\rho l_1}{I\rho l_2} = \frac{l_1}{l_2}. \qquad ...(1)$$

If one of the cells say ξ_2 is known (a standard cell) the emf of the other can be known. The equation (1) shows that for a steady current passing through the potentiometer wire, the potential difference across any length is proportional to the length of the wire.

CALIBRATION OF AN AMMETER AND VOLTMETER

The principle of standardisation is employed for the calibration of an ammeter and voltmeter. Connections are made. First the potentiometer wire is calibrated in milli-volt by obtaining the balance point with the standard cell of emf 1.0183 volts at 1018.3 cm. The balance point is obtained by varying Rh_1. Next with the help of Rh_2, the current in the ammeter A is adjusted and corresponding to this value of current the balance point on the potentiometer is noted. Then the current is calculated *i.e.,* directly the reading is current in amp. when multiplied by 10^{-3}. Similarly the voltmeter gives the voltage. The experiment is repeated for various readings of the ammeter and voltmeter. A calibration graph between the observed and calculated values is drawn.

MEASUREMENT OF VERY SMALL THERMO EMF

The emf of the order of 30 or 40 milli-volts e.q. thermo emf can be measured with potentiometer. First the potentiometer wire is calibrated in milli-volts with standard cell as discussed above. The connection are made. Let us take copper-iron couple to produce the thermo emf. As the potentiometer wire is calibrated *i.e.,* the p.d. per mm of the scale = $10^{-3} \times 0.001$ volt = 1 microvolt/mm = 1 milli-volt/cm. We can read directly the thermo-emf generated at various temperatures.

MEASUREMENT OF LARGE POTENTIAL DIFFERENCE

Though, the potentiometer is best suited for measuring the potential difference of the order of less than 2 volts, we can measure large potential difference by dividing it into fractions by volt box. A volt box consists of a very large resistance, about 10,000 ohms. This resistance is divided in fractions by intermediate tappings. The large potential to be measured is applied to this resistance.

Take any fraction, say 1/100 and balance it against the potentiometer wire and the p.d. for this fraction is known, then multiply it by 100 to obtain the requisite high potential.

DETERMINATION OF INTERNAL RESISTANCE OF A CELL

First the key K is open and the emf, ξ of the cell is balanced, say l_1 is the balancing length of the potentiometer wire. Now close the key K and introduce the resistance in the box R and the potential drop V = IR is balanced against a length l_2 of the potentiometer wire. If r is the internal resistance of the cell, then,

$$\frac{\xi}{V} = \frac{I(R+r)}{IR}$$

$$= \frac{R+r}{r} = \frac{l_1}{l_2}$$

or $$r = \frac{l_1 - l_2}{l_2} R. \qquad ...(1)$$

Hence r can be calculated.

COMPARISON OF TWO LOW RESISTANCES

Two low resistances of the same order can be compared using a potentiometer. The connections are made. As the same current is flowing through R_1 and R_2, then potential drops are IR_1 and IR_2 respectively. If the balancing lengths on the potentiometer wire are l_1 and l_2 respectively, then,

$$\frac{IR_1}{IR_2} = \frac{R_1}{R_2} = \frac{l_1}{l_2} \qquad ...(1)$$

DETERMINATION OF A GIVEN LOW RESISTANCE

If in the above experiment one of the resistance (R_2), is standard resistance then the other (R_1) can be calculated by relation

$$R_1 = R_2 \frac{l_1}{l_2}. \qquad ...(1)$$

Types of Potentiometers

Many different types of compact forms have been designed. The principle of working of all types of potentiometers is the same. The commonly used potentiometers are : (i) Crompton (ii) Student and (iii) Rayleigh potentiometer.

Crompton Potentiometer

Since it is difficult to get a very long, uniform homogenous wire, hence it is divided equally into fifteen segments. One of the wires is stretched on a wooden board and rest of the fourteen segments are repeated by the coils connected in series with the stretched wire, the resistance of each of coil is equal to the wire.

Student Potentiometer

This type of potentiometer was designed by Dr. N.K. Sethi. In this potentiometer, a platinum wire of length 50 cm and of uniform cross-section is used. Its resistance is 1 ohm. There are ten coils each of one ohm resistance put in series with the wire and on the scale each cm. is indicated as two.

Standardisation of Potentiometer of Direct Reading Potentiometer

A standard cell (cadmium cell) is connected the resistance is adjusted to 1018.3 ohm and the variable resistance (rheostat) so adjusted as to give no deflection with the standard cadmium cell in circuit. The current will be given by,

$$\frac{1.0183}{1018.3} = \times 10^{-3} \text{ amp},$$

where 1.0183 is the emf of the cadmium cell in volt. Now if the resistance with AB = 1 ohm as in the case of student potentiometer, then the potential difference per mm of the scale,

$$= 10^{-3} \times 0.001 \text{ volt/mm}$$

$$= 1 \text{ micro volt/mm} = 1 \text{ mili. volt/cm}$$

When the wire is calibrated to read one millivolt per cm the potentiometer will read the current in milliamperes per cm. *Thus, we can read current and voltage directly on the scale and such device is also called " direct reading" Potentiometer.*

Metal Electrode in an Electrolyte

When an electrode of pure zinc is immersed in an electrolyte, the chemical forces cause zinc ions to enter the liquid. Since zinc ions are positively charged, the electrode acquires a negative charge. *This process continues until the electric field due to the static charge on the electrode becomes great enough to counteract the chemical forces. Thus, when the* zinc ions are in electrostatics equilibrium, we have,

$$= \vec{E}_c + \vec{E}_{ch} \ 0. \qquad ...(2)$$

where $\vec{E}_c$ is the electric field intensity due to the static charge on the electrode and $\vec{E}_{ch}$ is the chemical force per unit charge.

Battery on Open Circuit

Positively charged copper ions tend to leave copper electrode just as positively charged zinc ions tent to leave the zinc electrode. The chemical forces may be considered as sum of two electric fields, $\vec{E}_{ch}$ acting on the surface of zinc and $\vec{E}_{ch}$ on the copper surface. $\vec{E}'_{ch}$ is greater in magnitude than $\vec{E}''_{ch}$ and the direction of $\vec{E}'_{ch}$ is opposite to that of $\vec{E}''_{ch}$. Thus, consider $\vec{E}_{ch}$, the chemical force per unit charge to be a vector function of position in the electrolyte, *i.e.*, resultant of these two, we have,

$$\vec{E}_{ch} = \vec{E}'_{ch} + \vec{E}''_{ch} \qquad ...(3)$$

The number of zinc ions entering the electrolyte is greater than copper ions. As a consequence the zinc electrode acquires a greater negative free charge than the copper electrode. Hence, *outside of the electrolyte, there is a field $\vec{E}_c$ due to the free charges on zinc and copper electrode $\vec{E}_c$ is directed from copper to zinc electrode.* In the electrolyte, a layer of positive ions surrounds each electrode. The electrode surface and the surrounding ions constitute a double layer of charges, the electrostatic filed intensity $\vec{E}_c$ is directed toward the electrode. When equilibrium is established in the double layer, we must have,

$$\vec{E}_c + \vec{E}_{ch} = 0 \qquad ...(4)$$

This equation must also be valid for points in the main body of the electrolyte where the ions are in electrostatic equilibrium. Otherwise there would be a continuous migration of ions. Since $\vec{E}_{ch}$ is zero at points relatively far from the electrode, $\vec{E}_c$ must be zero in the main body of electrolyte.

Definition of EMF

The *electromotive force (EMF) represents the work done by the non-electrostatic (chemical) forces during the transfer of a unit charge from the negative to the positive terminal.* The emf ξ of an electric cell is given by,

$$\xi = \int_m^p \vec{E}_{ch} \cdot d\vec{L} \qquad ...(5)$$

Since the electric field, $\vec{E}_{ch}$ exists only in the electrolyte, equation (5) may be written, here m denotes minus terminal and p positive,

$$\xi = \oint \vec{E}_{ch} \cdot d\vec{L} \qquad ...(6)$$

where the closed path extends from m to p through the electrolyte and then back to p through the space above the electrolyte. *No contribution to the final result is obtained by integrating from p to m because E_{ch} is zero in the space above the electrolyte.* According to equation (6), $\vec{E}_{ch}$ is "*non-conservative*", *because the line integral of* $\vec{E}_{ch}$ *around a closed path is not zero.*

DEFINITION OF POTENTIAL DIFFERENCE

Since there are now two electric fields, $\vec{E}_c$ and $\vec{E}_{ch}$, we must re-examine the definition of potential difference. The field $\vec{E}_c$ is due to the time dependent (static) charge distribution in the cell and is, therefore, "*conservative*" *and the potential difference is independent of path between the two points and can be computed only from the conservative electric field.* So the potential difference between the two plates is given by,

$$V_p - V_m = -\int_m^p \vec{E}_c . d\vec{L} \qquad ...(1)$$

Since the cell is on open circuit,

$$\vec{E}_c + \vec{E}_{ch} = 0$$

Therefore, substituting for $\vec{E}_c$, we have,

$$V_p - V_m = +\int_m^p \vec{E}_{ch} \cdot d\vec{L} \qquad ...(2)$$

Now according the right hand side of emf. Therefore,

$$V_p - V_m = \int_m^p \vec{E}_{ch.} d\vec{L} = \xi \qquad ...(3)$$

Thus, the potential, difference across a cell on open circuit is equal to the emf of the cell.

Since $\oint \vec{E}_c \cdot d\vec{L} = 0$, we can rewrite as follows,

$$\xi = \oint \left(\vec{E}_c + \vec{E}_{ch}\right) \cdot d\vec{L} \qquad ...(4)$$

The emf of an electric cell is equal to the "line integral" of the resultant electric field around a closed path that extends from the minus terminal to the plus terminal through electrolyte and then returns to the minus terminal through the space above the electrolyte.

Battery in Close Circuit

When an electric cell is connected to a resistance as in Fig. 2.3, the negative charge flows through the resistance from the electrode at lower potential (m) to the electrode at higher potential (p). The effect, of course, is the same as that of positive charge flowing from p to m.

Here, for *simplicity and algebraic consistency,* we assume that only positive charge is moving through a wire. When the negative charge on m is reduced. additional positive ions from m enter the solution. This means that in the vicinity of m, $\left|\vec{E}_{ch}\right| > \left|\vec{E}_c\right|$. When the negative charge on p is increased, practically no positive ions leave p to enter the solution. This means that in the vicinity of p,

$$\left|\vec{E}_c\right| > \left|\vec{E}_{ch}\right|.$$

The positive ions diffuse towards p and negative ions towards m. The current in the electrolyte consists of both of these streams of charges. The electric forces are such that positive charges is driven through the electrolyte from m to p. The charge returns to m through the external circuit because of the field $\vec{E}_c$. In a well-constructed cell, this process may continue for a long time.

Now the movement of charge in the electrolyte is related to the electric fields, $\vec{E}_c$ and $\vec{E}_{ch}$. We introduce the symbol $\vec{j}$ called current density which actually related to motion of charge in a conductor is proportional to the field or $\vec{j}$ is function of E,

$$\vec{J} = f(\vec{E}),\ J = 0 = f(0) \qquad ...(5)$$

Equation applies to ordinary conductors and excludes "superconductors" for which f(0) is a constant, for many conductors f is linear.

$$\vec{J} = \sigma\vec{E} \qquad ...(6)$$

In this case,

$$\vec{J} \propto (\vec{E}_c + \vec{E}_{ch})$$

$$= \sigma(\vec{E}_c + \vec{E}_{ch}) \qquad ...(7)$$

where the constant s is called "conductivity" Note that when the cell is disconnected from the resistor, $\vec{J} = 0$, hence, $\vec{E}_c + \vec{E}_{ch} = 0$.

This is same as open circuit case discussed previously.

CURRENT AND CURRENT DENSITY

If a net charge q passes through any cross-section of the conductor in time t, the current i is given by,

$$i = q/t \qquad ...(1)$$

Unit

The appropriate RMKS unit is ampere.

$$1 \text{ Ampere} = \frac{1\text{Coul}}{1\text{Sec.}}$$

The current given by equation (1) is the constant for the time. If the rate of flow of charge with time is not constant, then i varies with time and is given by,

$$i = \frac{dq}{dt} \qquad ...(2)$$

Here we shall consider only constant currents.

CURRENT DENSITY

Current i is a "characteristic" of a particular conductor. Current is a macroscopic quantity, like mass, volume, or length.

A *related* microscopic quantity is the *current density* $\vec{J}$. It is a vector and is *characteristic of a point inside a conductor rather than conductor as a whole*. If the current is distributed uniformly across a conductor or cross-sectional area A, the magnitude of current density for all points on the cross-section is,

$$J = i/A \qquad ...(1)$$

If the current is *not uniformly distributed* then equation (1) gives the *average density.* However, it is often of interest to consider the *current density at a point. This is defined as the current* Δ *i through a small area* Δ *s divided by* Δ *s, with the limit of this ratio taken as* Δ *s approaches zero.* Hence, the current density at a point is given by,

$$\vec{J} = \lim_{\Delta s \to 0} \frac{\Delta i}{\Delta s} \qquad ...(2)$$

Unit

Its unit in RMKS is ampere per square metre, as evident from equation (1).

The vector $\vec{J}$ at any point is oriented in the direction that a positive charge carrier would move at that point. An electron at that point would move in the direction $-\vec{J}$.

The relationship given in equation (2) can be *generalised* for any surface, for a particular surface in a conductor, i is the flux of the vector $\vec{J}$ over that surface or in other words, the *charge flow is represents by a vector* $\vec{J}$ *which gives the amount of charge passing per unit area per unit time through a surface element at right angles to the flow.*

If we consider the small area Δs given place in the material through which the charges are flowing, then amount of charge flowing across that area in a unit time is,

$$\Delta q = \vec{J}.\hat{n}\,\Delta s, \qquad ...(3)$$

where $\hat{n}$ is unit vector normal to Ds. Hence, the total charge passing per unit time through any surface,

$$q = \vec{J}.\int_8 \vec{J}.d\vec{S} \qquad ...(4)$$

From the definition of the current i = q/t, for unit time *i.e.*, t = 1, therefore,

$$i = q = \vec{J}.\int_8 \vec{J}.d\vec{S} \qquad ...(5)$$

Sensitivity of a Wheatstone's Bridge

The condition for the balance of Wheatstone's bridge is P/Q = R/S. This corresponds to zero current in the galvanometer. Now, if the bridge is slightly out of balance, there will be a current in the galvanometer, but the current is too feeble, hence cannot be detected easily so that the bridge may not be very sensitive. Thus, the sensitivities of the bridge depends on the sensitivity of the galvanometer.

In Wheatstone bridge, if we interchange the galvanometer and battery, the condition for the null point will not change, however, the two arrangements will not have the same sensitivity. The following discussion about the sensitiveness is due to H.A. Calendar.

The unknown resistance r is connected in the arm CD and i be the current in the arm CD and i be the current in it. Let the resistance in the arm AB, BC, DA and the galvanometer be mmr_1, mr_1, nr_1 and G respectively. Let the current in the galvanometer be i_g and in the arm AB i_1, then the current in the arm AB i_1, then the current in the arm AD will be $i_1 + i_g$. The bridge balance condition is according,

$$\frac{mnr_1}{mr_1} = \frac{nr_1}{r}$$

or $\qquad r_1 = r$

Thus, the bridge condition is satisfied if r1 = r or in other words, if $r = r_1$, the bridge is balanced. Hence, *the difference $r - r_1$ is the measure of want of balance i.e., the greater the difference, the bridge id far from the balance.* However, a bridge will be most sensitive, if the deflection in the galvanometer is large, even when $r - r_1$ is extremely small. Applying Kirchhoff's second law to the circuits ABDA and DCBD,

$$i_1mnr_1 - i_gG - (i + i_g)\, nr_1 = 0$$

and $ir - (i_1 + i_g)\, mr_1 - i_gG = 0$.

On rewriting the above equations,

$$i_1 mnr_1 - i_g (G + nr_1) - inr_1 = 0 \qquad ...(6)$$

and $-i_1mr_1 - i_g (G + mr_1) + ir = 0$ ...(7)

Multiplying equation (7) by n and adding the resulting equation to (6), we get,

$$in\,(r - r_1) - i_g\,[n(G + mr_1) + (G + nr_1)] = 0$$

or
$$\frac{i_g}{i} = \frac{n(r - r_1)}{(1 + n)G + n(1 + m)r_1}$$

or
$$\frac{i_g}{i} = \frac{r - r_1}{\left(1 + \frac{1}{n}\right)G + (1 + m)r_1} \qquad ...(8)$$

The ratio ig/i depends on the want of balance $r - r_1$, the galvanometer resistance G and the ratio m and n. This ratio is independent of the resistance of the battery.

For greater sensitiveness for a given want of balance, $r - r_1$, the value of i_g should be large, therefore,

(i) i should be large,

(ii) G should be small

(iii) n should be large

(iv) m should be small.

For increasing the sensitivity the current i should be large. But the current i cannot be increased indefinitely as this will heat up the resistance and consequently alter their resistance. Further, the sensitivity increases with decrease in the resistance of the galvanometer, but this decrease in the galvanometer resistance is achieved by decreasing the number of turns in the coil of the galvanometer and which *decreases the sensitivity of the galvanometer itself.*

Therefore, we have to increases sensitivity by increasing n and decreasing m. In the limiting case, if $n = \infty$ and $m = 0$,

$$\frac{i_g}{i} = \frac{r - r_1}{G + r_1} \qquad ...(9)$$

This condition gives the "*ideal sensitivity*" which is *impracticable* because (1) if n = ∞, the resistance of the arms AB and AD becomes infinite and (2) if m = 0; the points A and C are short circuited. However, making n very large and m very small has no great advantage, because even if n = m = 1, the sensitivity from equation (8) is

$$\frac{i_g}{i} = \frac{r - r_1}{2(G + r_1)} \qquad ...(10)$$

This value is still half of the value in the ideal limiting case given in equation (9).

Hence, the necessary condition is that n > 1 and m < 1 *i.e.*, n should not be very small or m too large because in either case the sensitivity will be very much reduced. Thus, *the Calender rule is that the sensitises of the bridge will be higher if the resistance in series with the unknown resistance is greater than the resistance connected in parallel to it.*

It can be easily shown that for greater sensitivity, the galvanometer should have the same order of the resistance as the unknown resistance. Another rule for greater sensitivensess is the "greater of the two resistances, the galvanometer or the battery should be connected between the junction of the two highest and two lowest resistances".

Maxwell's Cyclic Currents

Equations obtained by the application of Kirchhoff's law are sometimes very tedious to solve in a complicated network. Maxwell suggested a simple device, called cyclic currents method, which simplifies the solution of problems. According to this method, a certain magnitude of current is assumed to flow in each mesh, all the cyclic currents being in the same direction *i.e.,* either clockwise or anti-clockwise. The actual current flowing in any branch will then be the difference between the cyclic currents of the meshes, if separate.

WHEATSTONE'S BRIDGE

Let x, y and z be the currents flowing in the clockwise direction through the meshes ABDA, BCDB and ADCA. Applying Kirchhoff's second law, respectively we get,

$$xP + (x - y) R_g + (x - z) R = 0 \qquad ...(1)$$

$$yQ + (y - z) S + (y - x) R_g = 0 \quad ...(2)$$

$$(z - x) R + (z - y) S + zr = 0 \quad ...(3)$$

Collecting the coefficient of x, y and z in equation (1) and (2), we have,

$$x(P + R + R_g) - yR_g - zR = 0$$

and $-xR_g + y (Q + S + R_g) - zS = 0$

Multiplying the first of these equations by S and second by R and subtracting, we get,

$$x[S(P + R + R_g) + RR_g] - y[SR_g + R(Q + S + R_g)] = 0$$

The current in the galvanometer = x − y. At the balance point this is zero, *i.e.,* $x - y = 0$

or $\quad x = y.$

Hence, we get,

$$S(P + R + R_g) + RR_g = SR_g + R(Q + S + R_g)$$

or $\quad SP = RQ$

hence, $\dfrac{P}{Q} = \dfrac{R}{S}$

which is the required condition for balanced Wheatstone's bridge.

KELVIN'S METHOD FOR DETERMINING THE RESISTANCE OF A GALVANOMETER

In this method the galvanometer is put in the unknown arm of the post office box which is a Wheatstone bridge. The points B and D are connected by a tapping key K.

Let the clockwise cyclic currents through the meshes, (1) ABDA, (2), BCDB and (3) ABCEA be x, y and z respectively. Let P, Q, R, S, G and B be the resistance of the arms AB, BC, AD, DC BD and AC respectively. Applying Kirchhoff's second law to the meshes ABDA and BCDA, we get,

$$xP + (x - y) G + (x - z) R = 0 \quad ...(1)$$

$$yQ + (y - z) S + (y - x) G = 0 \quad ...(2)$$

The values of resistance P, Q and R are adjusted such that there is no charge in the deflection of the galvanometer when the key K_B is opened or closed.

This means that current in this arm must be zero *i.e.,*

$$x - y = 0$$

or $$x = y,$$

Substituting in equation (1) and (2), we get,

$$yP + (x - z)\,R = 0$$

$$yQ + (y - z)\,S = 0$$

or $$x(P + R) = zS$$

and $$y\,(Q + S) = zR$$

or ratio, $$\frac{x(P+R)}{y(Q+S)} = \frac{R}{S}$$

or $$\frac{P+R}{Q+S} = \frac{R}{S} \text{ since } x = y$$

$$PS + RS = RQ + RS$$

or $$S = R\,\frac{Q}{P} \qquad \text{...(3)}$$

Thus from the known values P, Q and R, S can be calculated.

Adjustment

(1) When the battery key K_B is pressed the galvanometer will always show a deflection as it is in one of the arms of the bridge. Generally, the deflection is out of scale, it may be reduced and brought within scale by connecting a high resistance box in the battery circuit.

(2) When the tapping key K_B is pressed the galvanometer deflection will change as some of the current will also flow in BD arm. P/Q is a fixed ratio for one observation, hence R is adjusted such that the galvanometer deflection remains the same before and after pressing key K_B.

In fact this method is a constant deflection method and not to be confused with null method. This method was first employed by Kelvin hence it is called Kelvin's method.

Mance's Method of Measuring the Internal Resistance of a Cell

Wheatstone's bridge circuit has been employed by Mance's to determine the internal resistance of the cell. The cell is connected in the unknown arm of the bridge and usual arm of the cell is connected by a key K_B. Let P, Q, R, r and G be the resistance of the arms AB, BC, DA, CD, and BD respectively.

Applying Kirchhoff's second law to the meshes ABDA and BCDB, we get,

$$(i - i_g - i_1) P - i_g G + (i - i_g) R = 0 \quad ...(4)$$

$$ir + i_g G + (i - i_1) Q = x \quad ...(5)$$

Collecting the coefficients of i, ig and i_1, we get,

$$i(P + R) - i_g(P + G + R) - i_1 P = 0 \quad ...(6)$$

$$i(r + Q) + i_g G - i_1 Q = x \quad ...(7)$$

Multiplying (3) by (r + Q) and (5) by (P + R), we get,

$$i(P + R)(r + Q) - i_g(P + G + R)(r + Q) - i_1 P(r + Q) = 0 \quad ...(8)$$

$$i(r + Q)(P + R) + i_g G(P + R) - i_1 Q(P + R) = x(P + R) \quad ...(9)$$

subtracting (5) from (6),

$$i_g[G(P+R) + (P+R)(r+Q)] - i_1\{Q(P+R) - P(r+Q)\} = x(P+R) \quad ...(10)$$

Adjustment

The values of P, Q, and R are adjusted as in the case of Kelvin's method (the procedure is exactly the same) such that the galvanometer shows a "constant deflection" when the key K_1 is pressed or opened.

That is the current in the galvanometer is constant, The equation (10) shows that ig will remain constant whatever be the value of current i1 if the coefficient of i_1 is zero.

$$Q(P + R) - P(r + Q) = 0$$

or $$Q(P + R) = P(r + Q)$$

$$PQ + QR = Pr + PQ$$

$$r = \frac{Q}{P} R \quad ...(11)$$

Hence r the internal resistance of the cell can be calculated by knowing P, Q and R.

Kelvin's Double Bridge for Low Resistance

The Wheatstone's bridge is not suitable for comparing two very low resistance such as metal rods because the junction resistances and resistances of connecting wires are larger compared to the low resistance to be measured. For accurate determination of low resistance, Kelvin devised double bridge.

PQ and RS are two low resistances of the same order of magnitude such as the *thick conducting rods of uniform cross-section.* One of which may be the known variable standard say PQ resistance and the other (RS) is unknown resistance.

The rods PQ and RS are connected in series through a mercury cup. The bridge is completed by two resistance coils AC and DF whose middle points are at B and E which are connected to the rods PQ and RS.

Let the resistance coil ABC be divided at B in the ratio r : mr and the coil DEF in the ratio r': nr'. Let the currents in various conductors be those indicated in the figure. If i' be the current between Q and R then at the junction Q the Kirchhoff's current law gives,

$$i_2 - i_3 - i' = 0$$

or $$i' = i_2 - i_3 \quad ...(12)$$

and at the junction R, we get,

$$i' + (i_3 + i_g) - i_4 = 0$$

or $$i' = i_4 - (i_3 + i_g) \quad ...(13)$$

Equating (12) and (13),

$$i_2 - i_3 = i_4 - (i_3 + i_g)$$

or $$i_g = i_4 - i_2 \quad ...(14)$$

Applying Kirchhoff's law to the meshes PQABEDP and RSFEBCR, we get,

$$i_2R_1 + i_3r - i_gG - i_1r' = 0 \quad ...(15)$$

$$i_4R_2 - (i_1 - i_g)\,nr' + i_gG + (i_3 + i_g)\,mr = 0 \quad ...(16)$$

The point P of the variable resistance which is in the form of knife edge so that it can slide along the rod is so adjusted that there is no deflection in the galvanometer

i.e., when $i_g = 0$,

then equation (15) and (16) become,

$$i_2R_1 + i_3r - i_1r' = 0 \quad ...(17)$$

$$i_4R_2 - i_1nr' + i_3nr = 0 \quad ...(18)$$

But from equation (14), when ig = 0

$$i_4 = i_2$$

therefore,

$$i_2R_1 + i_3r - i_2r' = 0$$

$$i_2R_2 - i_1nr' + i_3mr = 0$$

or $$i_2R_1 = i_2r' - i_3r$$

$$i_2R_2 = i_1nr' - i_3mr$$

or $$\frac{R_2}{R_1} = \frac{i_1nr' - i_3nr}{i_1r' - i_3r} \quad ...(19)$$

If the points B and E divide the two coils ABC and DEF in the same ratio, *i.e.*, when m = n, then,

$$\frac{R_2}{R_1} = m = n \quad ...(20)$$

Usually, we make,

$$m = n = 1$$

then, $R_2 = R_1$.

Thus, by suitably adjusting the variable point P of the known resistance, R_1 for zero deflection in the galvanometer, we get directly the unknown resistance R_2.

This bridge is suitable to measure resistances from 0.1 to 0.003 ohm.

Drift Speed (v_d) of Charge Carriers in a Conductor

The drift speed v_d of charge carriers in a conductor can be computed from the current density $\vec{J}$. Suppose we have a charge distribution, the conduction electrons (or charge) are moving to right with the constant drift speed v_d.

As this charge distribution passes over the surface elements Δs, the charge Δq flowing across this surface element in a time Δt is equal to the charge contained in a parallelepiped with Δs as its base and height $vd\Delta t$. If ρ is the volume density of the charge, then charge contained in the parallelepiped will be given by,

Charge passing through Δs in unit time

$$= \text{volume of parallelepiped} \times \rho$$

$$= \text{projection of } \Delta s \text{ perpendicular to } v_d \times \text{height of parallelepiped} \times \rho$$

$$\Delta q = \Delta s \cos\theta \; v_d \, \Delta \, t\rho$$

or $$\Delta q = \rho\vec{v}d.\vec{n}dS \;\Delta t$$

$$\text{Charge per unit time} = \rho\vec{v}d.\vec{n}dS = \rho\vec{v}d.d\vec{S} \quad ...(21)$$

Comparing equations (21) we get,

$$\vec{J} = \rho\vec{v}d \quad ...(22)$$

Let us consider the simple charge distribution of electrons each of charge e and N are the number of electrons per unit volume,

then,

$$r = Ne, \text{ hence, } \vec{J} = Ne\vec{v}d$$

or $$\vec{v}d = \vec{J}/Ne \quad ...(23)$$

Example:

Calculate the "drift speed" of electrons shown in Fig. 2.6 and its value when $\vec{J}$ *and* $\hat{n}$ *are parallel.*

Solution:

We shell use

$$i = \int_8 \vec{J}.d\vec{S}$$

Here, the value of integral is independent of the shape of the surface.

and, $i = \vec{J}.\vec{A}$

where, A is the area of cross-section of cylinder.

If area A is normal to the surface,

$i = J.A$, But $J = Nev_d$

Hence,

$$i = Nev_d A$$

or $$v_d = \frac{i}{NA_e}$$

Conservation of Charge *i.e.*, Continuity Equation

As we know that the electric charge is "*indestructible*", it is never *lost* or *created* that is the charge is conserved. The current i out of a closed surface represents the rate at which the charge leaves the volume v enclosed by 8. If there is a "*net current*" out of a closed surface *i.e.*, the net charge has crossed the surface. *The amount of charge inside must decrease correspondingly according to conservation rule,*

$$i = \int_8 \vec{J}.\hat{n}d\vec{S} = -\frac{d}{dt}(Q_{in}) \qquad ...(24)$$

where Q_{in} is the charge inside the surface. But

$$Q_{in} = \int_v \rho dv$$

therefore,

$$i = \int_S \vec{J}.\hat{n}d\vec{S} = -\frac{d}{dt}\int_v \rho dv$$

using divergence theorem, we get,

$$\int_8 \vec{J}.\hat{n}dS = \int_8 \vec{\nabla}.\hat{n}dS \; dv$$

hence $$\int_v \vec{\nabla}.\vec{J}.dv = -\frac{\delta}{\delta t}\int_v \rho dv$$

$$i.e., \ \nabla.\vec{J} = -\frac{\partial\rho}{\delta t} \qquad ...(25)$$

This is an important equation regarding "*charge flow*", *called continuity equation and describes the fact that the charge is conserved.* The decrease in value of charge inside a small volume must correspond to a flow of charge.

A steady state is one in which all time derivatives are zero so we must have $\frac{\delta\rho}{\delta t} = 0$ and

$$\vec{\nabla}.\vec{J} = 0 \text{ (Steady state)}$$

or $$\int_8 \vec{J}.d\vec{S} = 0 \qquad ...(26)$$

Now, if the medium obeys Ohm's law, we have,

$$\vec{J} = \sigma\vec{E}$$

then we get,

$$\vec{\nabla}.\vec{J} = \sigma\ \vec{\nabla}.\vec{E} = 0$$

So for a steady state we *conclude that E has zero divergence, hence according to Gauss' law, ρ vanishes and Laplace equation is valid.*

Since,

$$\vec{E} = -\vec{\nabla}.V,$$

we have Laplace equation,

$$-\vec{\nabla}^2 V = \vec{\nabla}.\vec{E} = \frac{\rho}{\varepsilon_0} = 0$$

Ohm's Law

Ohm's law states that the potential difference between the ends of a conductor varies directly as the current flowing in it, provided the temperature does not change and the physical state of the conductor remains the same.

Potential difference ∝ Current

or $$\frac{\text{Potential difference}}{\text{Current}} = \text{Constant}$$

= Resistance of the conductor.

Resistance

By definition, the resistance R_{ab} between the terminals a and b on a conductor is given by,

$$R_{ab} = \frac{V_a - V_b}{i} \quad ...(27)$$

where the current i is directed from a to b. This relation is called Ohm's law.

Unit

Resistance Rab is expressed in ohms

1 ohm = 1 Volt/1 ampere.

The ohm *is defined as the resistance of a conductor in which a potential difference of* 1 *volt is developed when a current of 1 ampere flows through it.*

Actually, *resistance has value beyond that of a mere definition.* For many conductors, the resistance depends only on the temperature and does not depend on the potential difference or the current. In *other words, the resistance is a property of the conductor and is not related to the circuit in which the conductor is connected.* The equation R = V/i can be used to calculate the current in a conductor if we know the resistance and the potential difference across the conductor.

Resistance of Cylindrical Conductor and Specific Resistance

An example let us calculate the resistance of the cylindrical conductor of radius r and length l.

The potential difference,

$$V_a - V_b = -\int_b^a \vec{E}.d\vec{l}$$

Now using $\vec{J} = \sigma\vec{E}$ and reversing the limit,

$$V_a - V_b = \int_a^b \frac{\vec{J}.d\vec{l}}{\sigma}$$

From definition, the magnitude of $J = \frac{i}{A} = \frac{i}{\pi r^2}$.

Hence,

$$V_a - V_b = \int_a^b \frac{idl}{\sigma \pi r^2}, \text{ (Since i and dl are parallel)}$$

$$= \frac{il}{\sigma \pi r^2}$$

Therefore,

$$R = \frac{V_a - V_b}{i} = \frac{1}{\sigma \pi r^2} \qquad ...(28)$$

The resistance of a conductor of uniform cross-section is directly proportional to the length and inversely proportional to the conductivity and the cross-sectional area.

SPECIFIC RESISTANCE

In relation (26), let us take a wire of unit length and unit cross-section, then,

$$R = \frac{1}{\sigma} = \rho_8 \qquad ...(1)$$

The quantity ρ_s is called "*specific resistance*" or "*resistivity*". Thus, the *specific resistance is the resistance offered by a conductor of unit length and unit cross-section.* Its unit is ohm-metre. "*Conductance,*" the "*reciprocal of resistivity*" is electrical conductivity or simply conductance. Its unit is mho.

OHM'S LAW AT A POINT

Consider a block of conducting material. Let a small imaginary rectangular shell of length l and cross-section a be constructed around a point P in the interior of the block with a normal to $\vec{J}$ as indicated.

Then, on applying Ohm's law to this shell, we have,

$$V = iR$$

where V = potential difference between ends of the shell.

But $V = E_l$, where E is the electric field and $J = i/a$

or $i = Ja$. Therefore,

$$E_l = V = iR = JaR$$

or $$\vec{J} = \frac{l}{aR} E$$

But according to equation (26), *i.e.*,

$$R = \frac{l}{\sigma a}$$

since here $\pi r^2 = a$,

or $$\sigma = \frac{l}{aR}$$

Therefore,

$$\vec{J} = \sigma\vec{E} \qquad ...(1)$$

By making the shell enclosing P as small as we wish, this relation may be made to apply at the point P and we may write as Eqn. (1).

Resistance Analogous to Reciprocal of Capacitance

In fact, this is the relationship between "conductivity" and "permittivity". In many respects, the calculation of the resistance between two electrodes is the same as that of capacitance. The similarity may be seen by comparing the definition of R and C, namely,

$$R = \frac{V_a - V_b}{i},\ C = \frac{V_a - V_b}{i}.$$

The current i is given by,

$$i = \oint_8 \vec{J}.d\vec{S}$$

Similarly, the charge q may be written by Gauss's law

$$q = \oint_8 \vec{D}.d\vec{S}$$

If the medium has a conductivity σ and permittivity ε, then

$\vec{J} = \sigma\vec{E}$ and $\vec{D} = \varepsilon\,\vec{E}$

Therefore,

$$i = \oint_8 \sigma\vec{E}.d\vec{S} \text{ and } q = \oint_8 \varepsilon\vec{E}.d\vec{S}$$

Thus,

$$R = \frac{V_a - V_b}{\oint_8 \sigma\vec{E}.d\vec{S}} \text{ and } C = \frac{\varepsilon\oint_8 \vec{E}.d\vec{S}}{V_a - V_b}$$

Consequently,

$$RC = \varepsilon/\sigma$$

$$R = \epsilon/\sigma C \qquad ...(2)$$

If the capacitance between two conductors separated by a medium of permittivity ε is known, then the resistance between the same two conductors separated by a medium of conductivity σ may be calculated by means of Eq. (2).

Example:

As an example, let us calculate the resistance between two parallel plates. The capacitance C between two parallel plates of area A separated by distance l is,

$$C = \varepsilon A/l$$

Solution:

By Eq. (2), the resistance,

$$R = \frac{\varepsilon}{\sigma(\varepsilon A/l)} = \frac{l}{\sigma A} = \frac{\rho_8 l}{A} \qquad ...(3)$$

Same result as expressed by and we can again define a specific resistance.

Wiedmann and Franz Law

As we know that good conductors of heat are also the good conductors for electric current. The Wiedmann and Franz *established "empirically" the relation between the thermal and electrical conductivities.* If σ_h is the thermal conductivity then the ratio σh/σ at higher temperature is proportional to the absolute temperature,

$$\frac{\sigma h}{\sigma} = LT \qquad ...(4)$$

where L is the constant of proportionality, is constant for all metals and is known as "Lorentz number".

THE RELAXATION TIME

The approach of current in an extended medium to a steady state can be studied by using,

$$\vec{\nabla}.\vec{J} + \frac{\delta\rho}{\delta t} = 0 \text{ and } \vec{J} = \sigma\vec{E}$$

together with $\vec{\nabla}.\vec{E} = \rho/\varepsilon o$, where ρ is charge density and we assume that the medium is conducting but not also polarizable Let us substitute $\vec{J} = \sigma\vec{E}$ in continuity equation, we get,

$$\vec{\nabla}\cdot\sigma\vec{E} + \frac{\delta\rho}{\delta t} = 0$$

or $$\sigma\vec{\nabla}\cdot\vec{E} + \frac{\delta\rho}{\delta t} = 0$$

or $$\frac{\sigma\rho}{\varepsilon_o} + \frac{\delta\rho}{\delta t} = 0, \text{ since } \nabla\cdot\vec{E} = \frac{\rho}{\varepsilon_o}$$

or $$\frac{\delta\rho}{\rho} = -\frac{\sigma dt}{\varepsilon_o}$$

Integrating the above equation,

$$\int\frac{d\rho}{\rho} = \log_e\rho = -\int\frac{\sigma}{\varepsilon_o}dt$$

$$= -\frac{\sigma t}{\varepsilon_o} + \text{Const (a)}$$

or $$\log_e\pi = -\frac{\sigma t}{\varepsilon_o} + a$$

Taking antilog, we get,

$$\rho = \exp\left[-\frac{\sigma t}{\varepsilon_o} + a\right]$$

or $= e^a \, e^{-\sigma t}/\varepsilon_o$

e^a can be evaluated by setting t = 0,

then, $\rho = \rho_o$, the fixed charge, *i.e.*,

$e^a = \rho_o$

hence, $\rho = \rho_o e^{-\sigma t}/\varepsilon_o$ ...(1)

We see that the time required for an initial charge density to fall to 1/e of its value is ε_o/σ. It is called "relaxation time".

Electric Power and Work

An electric generator connected to an electric heater. Let us calculate the amount of power supplied to the resistance R and the amount of heat developed in the resistor in a time t. The potential difference,

$$V_a - V_b = -\int_b^a \vec{E}.d\vec{l} = \int_a^b \vec{E}.d\vec{l}$$

The work done in moving a charge dq from a to b is,

$dw = (V_a - V_b)\, dq$ ...(2)

(Joules) = (Volts) (Coulombs)

By definition, power P is the rate of doing work. Thus,

$$P = \frac{dw}{dt} = (V_a - V_b)\frac{dq}{dt}$$

or $P = (V_a - V_b)\, i$

(Watts) = (Volts) (Amperes)

Since, by Ohm's law,

$V_a - V_b = iR$

hence,

P = i2R joules/see or watt

or, $$P = \frac{(V_a Vb)^2}{R} \text{ jouls/see. or watt} \quad ...(3)$$

This is the amount of energy absorbed per second in the resistance R. Total heat developed in R in time T,

$$w = \int_O^T Pdt = \int_O^T e^2 Rdt \quad ...(4)$$

where w is expressed in joules, equation (2) indicates that joule = watt. see.

EMF and Internal resistance

If a source of emf is connected in series with resistance R. If r is the internal resistance of B, the work done by the generator is the work done by the field E causing the charge to move from a to b.

Consequently the rate at which the generator does work on the battery is given by,

$$V_b - V_a = -\xi - ir$$

where, – x is the drop in voltage (due to emf) in passing from the plus terminal to the minus terminal and – ir is the drop in voltage in passing through the internal resistance r.

The generator maintains the potential difference $V_a - V_b$. The work done by the generator is the work done by the field E in moving charge from a to b. Consequently, the rate at which the generator does work on the battery is,

$$(V_a - V_b)\, i = \xi i + i^2 r$$

That the power absorbed in the battery due to emf is ξi and the power absorbed in the battery due to the evolution of heat is $i^2 r$. Finally, let us take the discharge of battery through the resistance.

$$\xi = iR + ir$$

multiply both sides of this equation by i, we get,

$$\xi i = i^2 R + i^2 r$$

Here ξi is the rate at which work is being done by chemical forces within the battery. This power is equal to the rate at which the heat is developed in the external resistance (i^2R) plus the rate at which heat is developed in the internal resistance (i^2r).

Maximum Power Theorem

Let the cell of emf ξ and internal resistance r be connected in series with an external resistance R. Then the current i in the circuit,

$$i = \frac{\xi}{R+r}$$

Power consumed $P = i^2R$

$$= \frac{\xi^2 R}{(R+r)^2}$$

$$= x^2R\ (R + r)^{-2}$$

Differentiating w.r.t. R,

$$\frac{dP}{dR} = x^2\ [(R + r)^{-2} - 2R\ (R + r)^{-3}]$$

$$= x^2\ (R + r)^{-3}\ [R + r - 2R]$$

$$= x^2\ (R + r)^{-3}\ (r - R)$$

$$\frac{dP}{dR} = \frac{\xi^2 (r-R)}{(R+r)^3}$$

Now, $\frac{dP}{dR} = 0$

when R = r or when R = ∞. If R = r, *the power consumed is maximum and if R = ∞ power consumed is zero.* This is known as "maximum power theorem" and *sates that the power consumed is maximum in the internal circuit if the external resistance is equal to the resistance of the source of emf.*

Theorem is applicable to cells, thermocouples, dynamos, thermionic valves etc.

Kirchhoff's laws

Ohm's law is unable to give values of currents in various parts of a complicated circuit. Kirchhoff in 1842 gave two general laws for calculating the currents and resistances in networks at junctions. *These laws are obtained from the law of "conservation of energy".*

FIRST CURRENT LAW OF KIRCHHOFF'S

It applies to circuit nodes (or junction) and states *that in any network, the algebraic sum of the currents at any junction in a circuit is zero, i.e.,*

$$\Sigma\, i = 0 \qquad ...(1)$$

the currents included in the above sum differ in sign according to whether they flow to or from the junction.

Proof:

We can prove this law as follows. For steady current the equation of continuity is given in equation (1), *i.e.,*

$$\int_8 \vec{J}.d\vec{S} = 0$$

and for a closed surface,

$$\oint \vec{J}.d\vec{S} = 0.$$

This relation can be applied to any volume. For example, the volume may be entirely inside of a conducting medium, or it may be only partially filled with conductors. The conductors may form a network inside the volume, or they may meet at a point.

As an illustration of this latter case, the surface S in Fig. 2.12 encloses a volume that contains five conductors meeting at a junction point P. *Taking the current flowing always from the junction as negative and the current flowing towards the junction as positive,*

according to relation $\oint \vec{J}d\vec{S} = 0.$

we get, $i_1 - i_2 - i_3 + i_5 = 0.$

It follows,

$$i_1 + i_4 = i_2 + i_3 + i_5.$$

In accordance with the above equation, Kirchhoff's current law can be stated in different way, *that the total current entering any junction (or node) of a circuit equals the total current leaving that junction.*

"Physically," Kirchhoff's current law *implies that there can be no "accumulation" of electric charge at any junction of a circuit and is nothing more than a restatement of the principle of conservation of charge.*

KIRCHHOFF'S VOLTAGE (SECOND) LAW OR LOOP RULE

It applies to close circuits (meshes) and states that in any closed circuit, *algebraic sum of the products of the current and resistance of the each part of the circuit is equal to the total emfs acting in the circuit, i.e.,*

$$\Sigma ir = \Sigma \xi \qquad ...(1)$$

and if ξ is not acting *i.e.,* if there is no source of emf, then,

$$\Sigma ir = 0$$

Proof:

To prove Kirchhoff's voltage law, let us consider the simple electric circuit. The current is i at all points in the circuit, we have from Ohm's law at a point,

$$\vec{J} = \sigma\vec{E}$$

where $\vec{E}$ is total field at the point.

In general the total $\vec{E}$ may not only be due to static charges but also to other causes such as the chemical action in a battery.

We may write,

$$\vec{E} = \vec{E}_c + \vec{E}_{ch}$$

where E_c is field due to static charge on the electrodes and E_{ch} is the field due to chemical forces.

If a is the cross-sectional area of the conductor, then,

$$J = i/a \text{ and } \sigma = l/aR$$

(See equation (2), where l is the length of the conductor).

Hence,

$$E = \frac{\vec{J}}{\sigma} = \frac{i}{a} = \frac{aR}{l} = \frac{iR}{l} = \vec{E} = \vec{E}_c + \vec{E}_{ch}$$

Integrating around the complete circuit, we get,

$$\oint \vec{E}_c . d\vec{l} + \oint \vec{E}_{ch} d\vec{l} = i \oint \frac{R}{l} dl$$

$\oint \vec{E}_c . d\vec{l} = 0$, because $\vec{E}_c$ is a conservative field and $\oint \vec{E}_{ch} . d\vec{l}$ is emf ξ according to equation (3), and $\oint R / l\, dl = R_T$. Hence,

where R_T is the total resistance and is equal to R if internal resistance is zero. We can generalize for number of closed circuits *i.e.,*

$$\Sigma\xi = \Sigma ir.$$

If ξ were absent, the,

$$\Sigma ir = 0.$$

While applying the second law, the *positive value of the current is taken when we are traversing in the direction of the current and the emf is taken positive when we traverse from the negative to the positive electrode through the electrolyte.* In circuit ABDA,

$$i_1 r_1 + i_2 r_2 + i_3 r_3 + ir = x$$

and in circuit ABCA

$$i_2 r_2 - i_5 r_5 - i_4 r_4 = 0.$$

Application of Kirchhoff's law to Wheatstone's Bridge

Represents the Wheatstone's bridge circuit where P, Q, R and S are connected to form a closed circuit. A cell is connected between the points A and C and a galvanometer is connected between the points B and D. The currents through the various branches are indicated in the figure. i_g is the current through galvanometer and G is its resistance. Applying the Kirchhoff's first law, we get, at the junction B,

$$i_1 - i_2 - i_g = 0 \qquad ...(4)$$

at the junction D,

$$i_3 + i_g - i_4 = 0 \qquad ...(5)$$

Now, applying Kirchhoff's second law to the meshes ABDA and ABCDA, we get,

$$i_1P + i_gG - i_3R = 0 \qquad ...(6)$$

$$i_1P + i_2Q - i_4S - i_3R = 0 \qquad ...(7)$$

When the galvanometer shows, "zero deflection", the points B and D are at the same potential and $i_g = 0$. Hence, equation (4), (5) and (6) reduce to

$$i_1 = i_2 \qquad ...(8)$$

$$i_3 = i_4 \qquad ...(9)$$

$$i_1P = i_3R \qquad ...(10)$$

Putting the values of (5) and (6) in equation (7), we get,

$$i_1P + i_1Q - i_3S - i_3R = 0$$

or $$i_1 (P + Q) = i_3 (R + S) \qquad ...(11)$$

Now divide equation (11) by (10), we get

$$\frac{i_1(P+Q)}{i_1P} = \frac{i_3(R+S)}{i_3R}$$

or $$\frac{P+Q}{P} = \frac{R+S}{R}$$

or $$\frac{Q}{P} = \frac{S}{R}$$

or $$\frac{P}{Q} = \frac{R}{S}. \qquad ...(12)$$

Thus, at balance point, *i.e., where the galvanometer shows zero deflection, we have, P/Q = R/S. We can find the value of an unknown resistance S in terms of known resistance P, Q and R.*

Galvanometer current i_g when the Wheatstone bridge is unbalanced. When the bridge is unbalanced, the galvanometer current, i_g can be determined by Kirchhoff's laws.

At the point A Kirchhoff's first law gives,

$$i - i_1 - i_3 = 0$$

or $$i_3 = i - i_1 \qquad ...(13)$$

At the point B, we get,

$$i_1 - i_2 - i_g = 0$$

or $$i_2 = i_1 - i_g \qquad ...(14)$$

and at the point C, we have,

$$i_2 + i_4 - i = 0$$

or $$i = i_2 + i_4 = i_1 - i_g + i_4$$

or $$i_4 = i - i_1 + i_g \qquad ...(15)$$

Applying Kirchhoff's second law to the mesh ABDA, we get,

$$i_1P + i_gG - i_3R = 0.$$

Substituting i3 from (1), we get,

$$i_1P + igG - (i - i1)\ R = 0$$

or $$i_1\ (P + R) + i_gG - iR = 0 \qquad ...(16)$$

From the mesh BCDB, we get,

$$i_2Q - i_4S - i_gG = 0.$$

Substituting i_2 and i_4 from (14) and (15) in the above equation, we get,

$$(i_1 - i_g)\ Q - (i - i_1 + i_g)\ S - i_gG = 0$$

or $$i_1\ (Q + S) - i_g\ (Q + S + G) - i_S = 0 \qquad ...(17)$$

Multiplying equation (16) by (Q + S) and (17) by (P + R), then subtracting resulting (17) from resulting (4), we get,

$$i_g\ [G(Q + S) + (P + R)\ (Q + S + G)] - i[R\ (Q+S) - S(P+R)] = 0$$

or $$i_g\ [G(Q + S) + (P + R)\ (Q + S + G)] - i(QR - SP) = 0$$

or $$i_g = i\,\frac{QR - SP}{G(P + Q + R + S) + (Q + S)(P + R)} \qquad ...(18)$$

From the above equation, we can also find the Wheatstone's bridge condition of balance by putting, ig = 0. Then,

$$QR - SP = 0$$

or $$P/Q = R/S$$

SOLVED EXAMPLES

Example 1:

In a neon gas discharge tube 2.9 × 10^{16} Ne + ions move to the right through a cross-section of the tube each second, while 1.2 × 10^{18} electrons move to the left in the same time. Find the magnitude and the direction of the current.

Solution:

The charge on each neon ion, q_1 = e 1.6 × 10^{-19} coul.

Let us take the positive direction from left to right and negative from right to left.

The current due to Ne + ions $= \dfrac{n_1 q_1}{t_1}$

$$= \frac{2.9 \times 10^{18} \times 1.6 \times 10^{-19}}{1} \text{ coul/sec.}$$

= 0.446 amp.

The charge on each electron, q_2 = – 1.6 × 10^{-19} coul.

The current due to electrons

$$= -\frac{n_2 q_2}{t_2} = -\frac{(1.2 \times 10^{18})(-1.6 \times 10^{-19})}{1}$$

= 1.2 × 1.6 coul/sec

= 1.92 amp.

Therefore, the net current through the discharge tube

= 0.446 + 1.92 = 2.366 amp.

Thus, the magnitude of the current is 2.366 amp. and its direction is from left to right.

Example 2:

An aluminium wire whose diameter is 0.245 cm. is welded end to end to a copper wire diameter 0.1626 cm. The composite wire carries a steady current of 10 amp.

(a) *What is the current density in each wire?*

(b) *Find the drift speed of the electrons in copper wire. Assume one free electron per atom in copper.*

Given Avogadro Number = 6 × 10^{23} atoms per gm-mole

Molecular weight of copper = 64

Density of copper = 9 gm/cc.

Solution:

(a) The current density is given by

$$\vec{j} = \frac{i}{A}.$$

The area of aluminium wire, $A_1 = \pi r_1^2$

$$= 3.14\left[\frac{0.245\times10^{-2}}{2}\right] = 4.713 \times 10^{-6}\ m^2$$

The area of copper wire, $A_2 = \pi r_2 2$

$$= 3.14\left[\frac{0.1626\times10^{-2}}{2}\right]^2 = 2.075 \times 10^{-6}\ m^2$$

Therefore, the current density in aluminium wire,

$$J_{Al} = \frac{i}{A_1} = \frac{10}{4.713\times10^{-6}} = 2.12 \times 10^{6}\ amp/m^2$$

similarly, the current density in copper wire,

$$J_{cu} = \frac{i}{A_2} = \frac{10}{2.075\times10^{-6}} = 4.818 \times 10^{-6}\ amp/m^2$$

(b) The number of atoms per unit volume = rN/M where r is the density of copper, N is Avogadro number and M is atomic weight of copper.

A given, there is one free electron per atom in copper, the number of free electrons per unit volume,

$$n = \frac{\rho N}{M}$$

$$= \frac{9\times(6\times10^{23})}{6}\ \text{electrons/cm}^3$$

$= 8.437 \times 10^{28}$ electrons/m^3

But, $J = nev_d$

Therefore

$$v_d = \frac{J_{cu}}{n_e} = \frac{4.818 \times 10^6}{8.437 \times 10^{23} \times 1.6 \times 10^{-19}}$$

$= 2.834 \times 10^{-4}$ m/sec.

Example 3:

The potential difference across a wire of 10^{-3} cm^2 cross-sectional area and 50 cm length is 2 volt, when a current of 0.25 amp. exists in the wire. Calculate,

(a) *The field strength in the wire,*

(b) *The current density, and*

(c) *The conductivity of the metal*

Solution:

(a) The field strength in the wire is given by,

$$E = \frac{V}{d} = \frac{2\,\text{volt}}{50\text{cm.}} = \frac{2\,\text{volt}}{0.5\text{metre}} = 4 \text{ volts/metre.}$$

(b) The current density,

$$J = \frac{i}{A} = \frac{0.25\text{amp}}{10^{-3}\,\text{cm}^2}$$

$$= \frac{0.25\text{amp.}}{10^{-3} \times 10^{-4}\,\text{m}^2}$$

$= 2.5 \times 10^6$ amp/m^2.

(c) The electrical conductivity of metal s is given by

$J = \sigma E$

or $$\sigma = \frac{J}{E} = \frac{(2.5 \times 10^6\,\text{amp./m}^2)}{(4\,\text{volts / metre})}$$

$= 6.25 \times 10^5$ mho/metre.

Example 4:

Calculate the insulation resistance of a cable having a cylindrical conductor of radius r_1 and the outer radius of insulation r_2. The specific resistance of the insulation is ρ.

Solution:

Let r and r + dr be the radii of thin concentric layer of insulation, then the surface area = 2πrl, where l is the length of the cable and

resistance of the layer = $\frac{\rho dr}{2\pi rl}$, according.

Therefore, total resistance of the cable of length l

$$R = \int_{r_1}^{r_2} \frac{\rho dr}{2\pi rl} = \frac{\rho}{2\pi l} \log_e \frac{r_2}{r_1}$$

Example 5:

Find the resistance of 1 cc of copper when drawn into a wire of diameter 0.32 mm specific resistance of copper = 1.59 × 10^{-6} ohms. cm

Solution:

Radius r of the wire = 0.016 cm

$$\text{Length of the wire} = \frac{\text{Volume}}{\pi r^2} = \frac{1}{3.14 \times 0.016^2}$$

and according to equation (2.30),

$$R = \frac{\rho l}{a} = \frac{1.59 \times 10^{-6}}{3.14 \times 0.016^2 \times 3.14 \times 0.016^2}$$

= 2.46 ohm.

Example 6:

A batt ry of 6 volts emf and 0.5 ohm internal resistance is joined in parallel with another of 10 volts emf and 1 ohm internal resistance. The combination is used to send a current through an external resistance. The combination is used to send a current through an external resistance of 12 ohm. Calculate, by application of Kirchhoff's law, the current through each battery.

Solution:

Let the current through the batteries B_1 and B_2 be I_1 and I_2 repulsively. Then applying Kirchhoff's first law to the point C, the current through the external resistance is $(I_1 + I_2)$.

Applying Kirchhoff's second law to the mesh ABFEA

$$I_1 \times 0.5 + (I_1 + I_2)\ 12 = 6$$

or $$12.5\ I_1 + 12\ I_2 = 6$$

$$25\ I_1 + 24\ I_2 = 12 \quad ...(1)$$

Similarly applying Kirchhoff's second law to the mesh CDFEC,

$$I_2 \times 1 + (I_1 + I_2)\ 12 = 10$$

or $$12\ I_1 + 13\ I_2 = 10 \quad ...(2)$$

Solving (1) and (2), we get,

$$I_1 = -\ 2.27 \text{ amp};\ I_2 = 2.865 \text{ amp}.$$

The negative sign in I_1 indicates that the current I_1 in the battery B is opposite to the direction.

Example 7:

12 wires, each of resistance r ohms, are connected in the form of a skeleton cube. Find the equivalent resistance of the cube when a cell is joined across any one of the wires forming the cube.

Solution:

Applying Kirchhoff's law to the mesh FCGHD and ABCDA,

$$zr - (y - z)\ r - 2\ (y - z)\ r - (y - z)\ r = 0$$

and $xr - yr - zr - yr = 0$

or $$5z - 4y = 0$$

or $$z = \frac{4}{5} \cdot y$$

and $x - 2y - z = 0$

or $$x - 2y - \frac{4}{5}y = 0$$

i.e., $y = \frac{5}{14}x.$

Let the equivalent resistance be R.

Then,

P.D. across AB = (x + 2y) R = xr.

Therefore,

$$R = \frac{x}{x+2y}r = \frac{x}{x+\frac{10}{14}x}r = \frac{7}{12}r$$

Example 8:

An aluminium wire whose diameter is 0.245 cm. is welded end to end to a copper wire diameter 0.1626 cm. The composite wire carries a steady current of 10 amp.

(a) *What is the current density in each wire?*

(b) *Find the drift speed of the electrons in copper wire. Assume one free electron per atom in copper.*

Given Avogadro Number = *6* × *10^{23} atoms per gm-mole*

Molecular weight of copper = *64*

Density of copper = *9 gm/cc.*

Solution:

(a) The current density is given by

$$\vec{J} = \frac{i}{A}.$$

The area of aluminium wire, $A_1 = \pi r_1^2$

$$= 3.14\left[\frac{0.245\times10^{-2}}{2}\right] = 4.713 \times 10^{-6}\ m^2$$

The area of copper wire, $A_2 = \pi r_2 2$

$$= 3.14\left[\frac{0.1626\times10^{-2}}{2}\right]^2 = 2.075 \times 10^{-6}\ \text{m}^2$$

Therefore, the current density in aluminium wire,

$$J_{Al} = \frac{i}{A_1} = \frac{10}{4.713\times10^{-6}} = 2.12 \times 10^{6}\ \text{amp/m}^2$$

similarly, the current density in copper wire,

$$J_{cu} = \frac{i}{A_2} = \frac{10}{2.075\times10^{-6}} = 4.818 \times 10^{-6}\ \text{amp/m}^2$$

(b) The number of atoms per unit volume = rN/M where r is the density of copper, N is Avogadro number and M is atomic weight of copper.

A given, there is one free electron per atom in copper, the number of free electrons per unit volume,

$$n = \frac{\rho N}{M}$$

$$= \frac{9\times(6\times10^{23})}{6}\ \text{electrons/cm}^3$$

$$= 8.437 \times 10^{28}\ \text{electrons/m}^3$$

But, $J = nev_d$

Therefore

$$v_d = \frac{J_{cu}}{n_e} = \frac{4.818\times10^{6}}{8.437\times10^{23}\times1.6\times10^{-19}}$$

$$= 2.834 \times 10^{-4}\ \text{m/sec.}$$

Example 9:

A source of emf ξ volts and internal resistance r ohms, supplies current to a heating coil. Calculate the resistance R of the heating coil in terms of r so that,

(1) *heating takes place most rapidly,*

(2) *3/4 of the total energy delivered by the source goes into the heating coil.*

Solution:

The current through the heating coil

$$i = \frac{\xi}{R+r} \qquad \text{...(1)}$$

The rate of energy delivered by the source to the heating coil,

$$W = i^2R = \left(\frac{\xi}{R+r}\right) R = \frac{\xi^2 R}{(R+r)^2} \qquad \text{...(2)}$$

(1) As we have discussed, for maximum power, the condition is that R = r.

(2) The rate of total energy delivered by the source,

$$= x_i = \frac{\xi^2}{R+r}$$

As given, $\frac{\xi^2 R}{(R+r)^2} = \frac{3}{4}\frac{\xi^2}{R+r}$

hence, after simplification, we get,

R = 3r.

Example 10:

A silver wire 1 mm in diameter carries a charge of 90 coulombs in 1 hour and 15 minutes. Silver contains 5.8 × 10^{22} free electrons/cm^2. Calculate, (i) the current in the wire in amperes, (ii) the drift velocity of the electrons in wire in metres/sec.

Solution:

(1) 1 hour and 15 minutes = 4500 secs.

Current i = q/t

i = 90/4500 = 0.02 ampere

(2) Current density $\vec{j} = \frac{\text{Current}}{\text{Area}}$

$$= \frac{0.02 \text{amp}}{\pi \times (0.0005)^2 \text{ m}^2}$$

$$= \frac{0.02}{3.14 \times 5 \times 5 \times 10^{-8}} \text{ amp/m}^2$$

$= 2.55 \times 10^4$ amp/m^2

The drift velocity,

$$v_d = \frac{J}{ne} = \frac{2.55 \times 10^4}{1.6 \times 10^{-19} \times 5.8 \times 10^{28}} \text{ m/sec.}$$

(Since 1 cm^3 = 10^{-6} m^3)

hence, n = 5.8×10^{28} elect.m^3

$v_d = 2.69 \times 10^{-6}$ metre/sec.

EXERCISES

1. Define emf and potential difference. What is the difference between an emf and potential difference?
2. What do you mean by electric current and current density? How are they related to each other? Establish the relation between current density and the velocity of charge carrier.
3. What is the conservation of charge rule? Show that

   $$\nabla.\vec{J} + \frac{\delta p}{\delta t} = 0.$$

4. Define (a) resistance, (b) resistivity, (c) conductivity. Establish the relation $\vec{J} = \sigma\vec{E}$.
5. Give Kelvin's method for measuring the galvanometer resistance.
6. Describe how you will measure the internal resistance of a cell using Wheatstone's bridge.
7. In a cloud of electrons having 6×10^{29} electrons per metre^3, 40 per cent of the electrons have a drift velocity 1 mm per second in x-direction, 40 per cent have drift velocity of 2 mm per second in y-direction and remaining 20 per cent have drift velocity of 3 mm per second in z-direction. Calculate the current density $\vec{J}$.
8. What is the relation between conductivity and permittivity? Show that R = $\varepsilon/\sigma c$.

9. State and prove Kirchhoff's laws of distribution of currents in an electrical network and apply these laws to deduce the condition of

   (a) Balance of Wheatstone's bridge.

   (b) Maximum sensitivity of Wheatstone Bridge.

10. Explain the Maxwell's cyclic currents method. Apply it to obtain the Wheatstone's bridge condition.

11. An aluminium wire of diameter 0.10 cm is welded end to ent to a copper wire of diameter 0.064 cm. The composite wire carries a steady current of 10 amp. What is the current density in each wire? What is the drift velocity of electrons in copper density in each wire? Given that the density of copper is 9.0 gms/se and Avogadro's number is 6×10^{23} atoms/mol.

12. (a) Let n resistances $R_1, R_2 \ldots R_n$ be connected in series. Show that the equivalent resistance R is given by,

$$R = \sum_{i=1}^{n} R_i$$

   (b) Let n resistances $R_1, R_2 \ldots R_n$ be connected in parallel. Show that the equivalent resistance R is given by,

$$\frac{1}{R} = \sum_{i=1}^{n} \frac{1}{R_i}.$$